Sobolev Spaces of Infinite Order and Differential Equations

Mathematics and Its Applications (East European Series)

Julij A. Dubinskij

Department of Applied Mathematics,
Moscow Power Engineering Institute, U.S.S.R.

# Sobolev Spaces of Infinite Order and Differential Equations

D. REIDEL PUBLISHING COMPANY

A MEMBER OF THE KLUWER ACADEMIC PUBLISHERS GROUP

DORDRECHT / BOSTON / LANCASTER / TOKYO

Library of Congress Cataloging-in-Publication Data

CIP

Dubinskiĭ, IU. A.

Sobolev spaces of infinite order and differential equations.

(Mathematics and its applications. East European series)

Bibliography: p.
Includes index.
1. Sobolev spaces. 2. Differential equations.
I. Title. II. Series: Mathematics and its applications
(D. Reidel Publishing Company). East European series.
QA323.D8 1986 515.7'3 85-24484

ISBN 90-277-2147-5

Distributors for the Socialist Countries
BSB B. G. Teubner Verlagsgesellschaft, Leipzig, G.D.R.

Distributors for the U.S.A. and Canada
Kluwer Academic Publishers,
190 Old Derby Street, Hingham, MA 02043, U.S.A.

Distributors for all remaining countries
Kluwer Academic Publishers Group,
P.O. Box 322, 3300 AH Dordrecht, Holland.

Published by BSB B. G. Teubner Verlagsgesellschaft, Leipzig, G.D.R.
in co-edition with
D. Reidel Publishing Company, Dordrecht, Holland.



Printed in the German Democratic Republic

# EDITOR'S PREFACE

Approach your problems from the right end and begin with the answers. Then one day, perhaps you will find the final question.

'The Hermit Clad in Crane Feathers' in R. van Gulik's *The Chinese Maze Murders*.

It isn't that they can't see the solution. It is that they can't see the problem.

G.K. Chesterton. *The Scandal of Father Brown* 'The point of a Pin'.

Growing specialization and diversification have brought a host of monographs and textbooks on increasingly specialized topics. However, the "tree" of knowledge of mathematics and related fields does not grow only by putting forth new branches. It also happens, quite often in fact, that branches which were thought to be completely disparate are suddenly seen to be related.

Further, the kind and level of sophistication of mathematics applied in various sciences has changed drastically in recent years: measure theory is used (non-trivially) in regional and theoretical economics; algebraic geometry interacts with physics; the Minkowsky lemma, coding theory and the structure of water meet one another in packing and covering theory; quantum fields, crystal defects and mathematical programming profit from homotopy theory; Lie algebras are relevant to filtering; and prediction and electrical engineering can use Stein spaces. And in addition to this there are such new emerging subdisciplines as "experimental mathematics", "CFD", "completely integrable systems", "chaos, synergetics and large-scale order", which are almost impossible to fit into the existing classification schemes. They draw upon widely different sections of mathematics. This programme, Mathematics and Its Applications, is devoted to new emerging (sub)disciplines and to such (new) interrelations as exempla gratia:

- a central concept which plays an important role in several different mathematical and/or scientific specialized areas;
- new applications of the results and ideas from one area of scientific endeavour into another;
- influences which the results, problems and concepts of one field of enquiry have and have had on the development of another.

The Mathematics and Its Applications programme tries to make available a careful selection of books which fit the philosophy outlined above. With such books, which are stimulating rather than definitive, intriguing rather than encyclopaedic, we hope to contribute something towards better communication among the practitioners in diversified fields.

Because of the wealth of scholarly research being undertaken in the Soviet Union, Eastern Europe, and Japan, it was decided to devote special attention to the work emanating from these particular regions. Thus it was decided to start three regional series under the umbrella of the main MIA programme.

There are good reasons to believe that in the future, looking back, the 20-th century will be regarded as the age of functional analysis or, more generally the period in which various stings were taken out of infinity. Thus we have, by and large, learned now to live with infinite dimensional (function) spaces and operators and functionals on them. There are however other finiteness aspects which should probably be removed both in view of applications and for greater power and elegance of theory. One such is that it is (historically) customary to restrict attention to differential operators of finite order. In this book, differential operators of infinite order are considered and the Sobolev space theory needed to live with them is developed and applied. It is probably unnecessary to argue that differential operators are important and occur naturally in many places. There is an increasing algebraic and formal manipulation aspect to analysis at the moment, and in my personal opinion $\infty$-order differential operators will among others play an important role in putting formal algebraic arguments with differential operators (and their inverses) on a sound footing.

The unreasonable effectiveness of mathematics in science ...

Eugene Wigner

Well, if you know of a better 'ole, go to it.

Bruce Bairnsfather

What is now proved was once only imagined.

William Blake

As long as algebra and geometry proceeded along separate paths, their advance was slow and their applications limited.

But when these sciences joined company they drew from each other fresh vitality and thenceforward marched on at a rapid pace towards perfection.

Joseph Louis Lagrange.

Bussum, July 1985

Michiel Hazewinkel

To my father -
a schoolteacher
of mathematics

## PREFACE

The present book is devoted to the study of boundary value problems of infinite order and the corresponding functional spaces. Despite the fact that the first results in this direction were obtained 10 years ago (see Bibl.), one can now suggest rather good foundations (of course, in our opinion) for the theory of both topics mentioned above.

Before each Chapter there is an introduction, so we do not describe here the contents in detail, but give only their general character. From this point of view the material of our book may be divided into two main parts:

1) the theory of boundary value problems of infinite order itself (Ch. II, III, VI);

2) the theory of Sobolev spaces of infinite order $W^{\infty}\{a_{\alpha}, p_{\alpha}\}$ which are the "energy" spaces of the corresponding problems (Ch. I, III - V).

Two theories are of primary importance in the development of the questions described in this book: the theory of nonlinear boundary value problems of finite order and the theory of classical Sobolev spaces $W_p^m$.

As is known, at present these two theories are essentially two sides of one theory of differential equations of finite order in the corresponding functional spaces. This is also true for the differential equations of infinite order. Moreover, in this case the connection between the boundary value problems and their functional spaces is deeper, since the existence (nontriviality) of energy spaces $W^{\infty}\{a_{\alpha}, p_{\alpha}\}$ itself is, essentially, equivalent to the correctness of the corresponding boundary value problems. In this connection we start with the study of the question of nontriviality of the necessary spaces. The settlement of this question of nontriviality allows for the investigation of not only the corresponding boundary value problems, but also a series of related functional problems: imbedding theory of $W^{\infty}\{a_{\alpha},p\}$, trace theory of functions $u(x) \in W^{\infty}$ $a_{\alpha},p_{\alpha}$, geometrical characteristics of these spaces etc., all of which are of independent interest.

The present book includes, in particular, the systematic consideration of these questions.

Moscow Energy Institute

Moscow 1984

CONTENTS

CHAPTER I

# NONTRIVIALITY OF SOBOLEV SPACES OF INFINITE ORDER

## Introduction

In this Chapter we introduce some functional spaces $W^{\infty}\{a_\alpha, p_\alpha\}$ the "metric" of which is determined by the series

$$\varrho(u) \equiv \sum_{|\alpha|=0}^{\infty} a_\alpha \|D^\alpha u\|_{r_\alpha}^{p_\alpha} < \infty ,$$

where $a_\alpha \geqq 0$, $p_\alpha \geqq 1$, $r_\alpha \geqq 1$ are arbitrary sequences of numbers; $\|\cdot\|_r$ is the norm in Lebesgue space $L_r$.

We call such spaces the Sobolev spaces of infinite order. Infinitely differentiable functions $u(x)\colon G \to \mathbb{C}^1$ $(G \subset \mathbb{R}^n)$ are the elements of these spaces; moreover, the functions $u(x)$ may satisfy some boundary conditions.

In contrast with the finite order Sobolev spaces, the very first question, which arises in the study of the spaces $W^{\infty}\{a_\alpha, p_\alpha\}$, is the question of their nontriviality (or nonemptiness), i. e. the question of the existence of a function $u(x)$ such that $\varrho(u) < \infty$

It turns out that the answer of this question depends not only on the given parameters $a_\alpha$, $p_\alpha$ of the space $W^{\infty}\{a_\alpha, p_\alpha\}$, but also on the region $G$ (the domain of definition of the functions $u(x) \in W^{\infty}\{a_\alpha, p_\alpha\}$), boundary values of the functions $u(x)$ etc.

In §§ 1 - 3 we give necessary and sufficient conditions for nontriviality of Sobolev spaces of infinite order in three of the most commonly encountered cases in analysis: a bounded region $G \subset \mathbb{R}^n$, full Euclidean space $\mathbb{R}^n$ and the torus $T^n = S^1 \times \ldots \times S^1$, where $S^1$ is the unit circle. In § 4 the case of the strip is considered.

In the case of a bounded region $G \subset \mathbb{R}^n$ the question of nontriviality of $\overset{\circ}{W}{}^{\infty}\{a_\alpha, p_\alpha\}$ turns out to be closely related to the classical theory of Hadamard's quasianalytic classes $C\{M_N\}$. Namely, the space $\overset{\circ}{W}{}^{\infty}\{a_\alpha, p_\alpha\}$ is nontrivial precisely when a certain sequence $M_N > 0$, which is uniquely defined by the numbers $a_\alpha$ and $p_\alpha$ for $|\alpha| = N$, generates a one-dimensional non-quasianalytic class $C\{M_N\}$.

In the cases of the torus and full Euclidean space $\mathbb{R}^n$ the nontriviality of the space $W^{\infty}\{a_\alpha, p_\alpha\}$ is connected with the characteristic function of this space

$$\varphi(\xi) \equiv \sum_{|\alpha|=0}^{\infty} a_\alpha \xi^{\alpha p_\alpha}, \quad \xi = (\xi_1, \dots, \xi_n).$$

It is very simple to formulate the criterion of nontriviality of $W^\infty\{a_\alpha, p_\alpha\}$ in the case of a bounded sequence $p_\alpha$ and, in particular, in the Banach case $p_\alpha \equiv p$. Namely, if $p_\alpha \leqq$ const, then $W^\infty\{a_\alpha, p_\alpha\}(T^n)$ is nontrivial if and only if the function

$$a(z) \equiv \sum_{|\alpha|=0}^{\infty} a_\alpha z^\alpha, \quad z \in \mathbb{C}^n,$$

is an entire function.

Under the same conditions the nontriviality of the space $W^\infty\{a_\alpha, p_\alpha\}(\mathbb{R}^n)$ of functions $u(x)\colon \mathbb{R}^n \to \mathbb{C}^1$ is equivalent to the analyticity of the function $a(z)$ in a neighbourhood of zero.

In the case of the strip $G = [0,a] \times \mathbb{R}^n$ the criterion for nontriviality of the space $\overset{\circ}{W}{}^\infty\{a_{n\alpha}, p_\alpha\}$ is, roughly speaking, a combination of the criterion for nontriviality of the space $\overset{\circ}{W}{}^\infty$ on the interval $(0,a)$ and the criterion for nontriviality of the space $W^\infty$ in the full Euclidean space.

## § 1. Criterion for nontriviality of the spaces $\overset{\circ}{W}{}^\infty\{a_\alpha, p_\alpha\}$ in the case of a bounded domain

Let $G \subset \mathbb{R}^n$ be a bounded domain and let $\Gamma$ be the boundary of this domain. Let us denote the space of infinitely differentiable functions $u(x)\colon G \to \mathbb{C}^1$ such that $D^\omega u|_\Gamma = 0$, $|\omega| = 0,1,\dots$, by $C_0^\infty(G)$. (Here $\omega = (\omega_1, \dots, \omega_n)$, $\omega_j \in \mathbb{N}$, $1 \leqq j \leqq n$; $D^\omega$ is the standard notation

$$D^\omega = \frac{\partial^{|\omega|}}{\partial x_1^{\omega_1} \dots \partial x_n^{\omega_n}}, \qquad |\omega| = \omega_1 + \dots + \omega_n;$$

in the same way $D^\alpha$, $D^\gamma$ etc.). In other words

$$C_0^\infty(G) = \left\{ u(x) \in C^\infty(G)\colon D^\omega u|_\Gamma = 0, \ |\omega| = 0,1,\dots \right\}.$$

One can suppose that $u(x) \equiv 0$ for $x \bar{\in} G$. We shall refer to such a function as a finite function.

Let us consider the, following functional space

$$\overset{\circ}{W}{}^{\infty}\{a_\alpha, p_\alpha\} = \left\{u(x) \in C_0^\infty(G) : \varrho(u) \equiv \sum_{|\alpha|=0} a_\alpha \|D^\alpha u\|_{r_\alpha}^{p_\alpha} < \infty\right.,$$

where $a_\alpha \geqq 0$, $p_\alpha \geqq 1$, $r_\alpha \geqq 1$ are arbitrary sequences of numbers; $\|\cdot\|_r$ is the norm in $L_r(G)$.[1)]

It is clear that the question of nontriviality of $\overset{\circ}{W}{}^{\infty}\{a_\alpha, p_\alpha\}$ arises if among the numbers $a_\alpha \geqq 0$ there are infinitely many greater than zero.

Definition 1.1. The space $\overset{\circ}{W}{}^{\infty}\{a_\alpha, p_\alpha\}$ is called nontrivial if it contains at least one function which not identically equal to zero, i. e. there is a function $u(x) \in C_0^\infty(G)$ such that $\varrho(u) < \infty$ .

Before formulating a nontriviality criterion let us introduce the following number sequence. Namely, let $M_N$, $N = 0,1,\dots$, be the solution of the equation

$$\sum_{|\alpha|=N} a_\alpha M_N^{p_\alpha} = 1 \tag{1.1}$$

with $M_N = +\infty$ if $a_\alpha \equiv 0$ for all $|\alpha| = N$. (Obviously, the relations (1.1) define the numbers $M_N$ uniquely.)

Theorem 1.1. The space $\overset{\circ}{W}{}^{\infty}\{a_\alpha, p_\alpha\}$ is nontrivial if and only if the sequence $M_N$, $N = 0,1,\dots$, defines a non-quasianalytic Hadamard's class of one real variable.

Remark. Let us reeall the definition of Hadamard's class $C\{M_N\}$, where $M_N$, $N = 0,1,\dots$, is a sumber sequence. Namely,

$$C\{M_N\} = \left\{u(x) \in C^\infty(a,b) : |D^N u(x)| \leqq KM_N, \quad x \in (a,b), \quad N = 0,1,\dots\right\},$$

where $K > 0$ is a constant, depending (in general) on the function $u(x)$. The class $C\{M_N\}$ is called quasianalytic when the following condition is valid:

$$\text{if } u(x) \in C\{M_N\},\ v(x) \in C\{M_N\} \text{ and } D^N u(x_0) = D^N v(x_0),\ N = 0,1,\dots,$$

($x_0 \in (a,b)$ is a fixed point), then $u(x) \equiv v(x)$ for all $x \in (a,b)$. Otherwise the class $C\{M_N\}$ is called non-quasianalytic.

---

1) As one can see below, the values $r_\alpha$ are immaterial in the question of nontriviality. In view of this fact we don't use $r_\alpha$ in the notation.

The various algebraic conditions (criteria) of non-quasianalyticity are well known (see, for example, S. MANDELBROJT [1]). One of them - the Mandelbrojt-Bang criterion - will be used for the proof of our theorem.

Proof of the Theorem 1.1. To prove the necessity of our conditions it is obviously enough to prove that if the sequence $M_N$ defines a quasianalytic Hadamard's class $C\{M_N\}$ and $u(x) \in \overset{\circ}{W}{}^{\infty}\{a_\alpha, p_\alpha\}$, then $u(x) \equiv 0$. Indeed, $u(x) \in \overset{\circ}{W}{}^{\infty}\{a_\alpha, p_\alpha\}$ implies $u(x) \in C_0^\infty(G)$ and, consequently, for any $\alpha$ and $\xi = (\xi_1, \ldots, \xi_n)$ one has the inequality

$$|\xi^\alpha| \cdot |\tilde{u}(\xi)| \leqq \|D^\alpha u\|_1 \leqq K \|D^\alpha u\|_{r_\alpha}, \tag{1.2}$$

where $\tilde{u}(\xi)$ is the Fourier transform of $u(x)$ and $K > 0$ is a constant, depending only on the measure of $G$.

Let $\xi = \eta\Theta$, where the vector $\Theta = (\Theta_1, \ldots, \Theta_n)$, $|\Theta_j| \geqq 1$, $1 \leqq j \leqq n$, is fixed and $\eta \in \mathbb{R}^1$ is arbitrary. Then, using (1.2), we obtain the inequality

$$|\eta|^N |\tilde{u}(\eta\Theta)| \leqq K \|D^\alpha u\|_{r_\alpha}, \quad |\alpha| = N.$$

Since the natural number $N$ is arbitrary and the function $u(x)$ is finite, we obtain that for any $N \geqq 2$

$$(1 + \eta^2) |\eta|^{N-2} |\tilde{u}(\eta\Theta)| \leqq K \|D^\alpha u\|_{r_\alpha}, \quad |\alpha| = N,$$

where $K > 0$ is a constant.[1)]

Thus,

$$\sum_{|\alpha|=N} a_\alpha \left[(1 + \eta^2) |\eta|^{N-2} |\tilde{u}(\eta\Theta)| K^{-1}\right] \leqq \sum_{|\alpha|=N} a_\alpha \|D^\alpha u\|_{r_\alpha}^{p_\alpha}.$$

Since $u(x) \in \overset{\circ}{W}{}^{\infty}\{a_\alpha, p_\alpha\}$, one can suppose (without less of generality) that

$$\sum_{|\alpha|=N} a_\alpha \|D^\alpha u\|_{r_\alpha}^{p_\alpha} \leqq 1$$

and, consequently, taking into account the definition of sequence $M_N$, $N = 0, 1, \ldots$, we shall have the inequalities

$$(1 + \eta^2) |\eta|^{N-2} |\tilde{u}(\eta\Theta)| \leqq K M_N, \quad N = 2, 3, \ldots$$

---

1) Here and below all constants the values of which are non-principal will be denoted by one letter K.

The latter inequalities mean that

$$|D^{N-2}v(y)| \leqq KM_N, \ N = 2,3,\ldots,$$

where the function $v(y)$, $y \in \mathbb{R}^1$, is the inverse Fourier transform of $\tilde{u}(\eta\Theta)$ with respect to variable $\eta \in \mathbb{R}^1$.

These inequalities imply that the function $v(y)$ belongs to the Hadamard class $C\{M_{N+2}\}$ which by assumption (together with $C\{M_N\}$) is quasianalytic. On the other hand, from the Paley-Wiener theorem the function $v(y) \in C_0^\infty(\mathbb{R}^1)$; consequently, $v(y) \equiv 0$. Thus for any line $\xi = \eta\Theta$, where $\Theta = (\Theta_1,\ldots,\Theta_n)$, $\Theta_j \geqq 1$, we obtain that $\tilde{u}(\xi) = 0$, i. e. $\tilde{u}(\xi) \equiv 0$ for all $\xi \in \mathbb{R}^n$. It follows that $u(x) \equiv 0$ in $G$. The necessity is proved.

Sufficiency. The known lemma on functions with compact support of one real variable plays a fundamental part in this proof (cf. S. MANDELBROJT [1], Ch. IV, Theorem 4.1.IV).

Lemma 1.1. Let $\mu_0 = 1$, $\mu_N > 0$ $(N = 1,2,\ldots)$ be a number sequence satisfying the condition

$$\mu_1 + \mu_2 + \ldots < \frac{a}{3}, \quad a > 0. \tag{1.3}$$

Then there exists a function $v(t) \in C_0^\infty(-a,a)$, $t \in \mathbb{R}^1$, such that:

1. $v(0) = 1$, $D^N v(-a) = D^N v(a) = 0$, $N = 0,1,\ldots;$
2. for any $t \in (-a,a)$

$$|D^N v(t)| \leqq (\mu_0\mu_1\cdots\mu_N)^{-1}, \quad N = 0,1,\ldots \tag{1.4}$$

Proof of lemma. We choose the continuous function $v_0(t)$ satisfying the following conditions:

1. $0 \leqq v_0(t) \leqq 1$, $t \in (-a,a)$;
2. $v_0(t) \equiv 1$, if $t \in (-a/3,a/3)$;
3. $v_0(t) \equiv 0$, if $t \in (-a,-2a/3)$ and $t \in (2a/3,a)$.

Besides let $v_0(t)$ be an even function.

Further we define the sequence of functions $v_m(t)$ by the following recursion formula

$$v_m(t) = \frac{1}{2\mu_m} \int_{t-\mu_m}^{t+\mu_m} v_{m-1}(\tau)d\tau, \quad m = 1,2,\ldots$$

In view of condition (1.3) these functions are finite on the interval $(-a,a)$. It is also clear that the functions $v_m(t)$, $m = 1,2,\ldots$, are even and differentiable at least up to order m.

Let us prove that the sequence $v_m(t)$ (strictly speaking a subsequence) converges to a function $v(t) \in C_0^\infty(-a,a)$ as $m \to \infty$. To establish this fact we shall prove first the following inequality

$$\max_{t\in(-a,a)} |D^m v_m(t)| \leqq (\mu_0\mu_1\ldots\mu_m)^{-1}, \quad m = 0,1,\ldots \tag{1.5}$$

In fact, for m = 0 the inequality (1.5) is evident. Further, for $n \leqq m$ we have

$$D^n v_m(t) = \frac{1}{2\mu_m}\left[D^{n-1} v_{m-1}(t+\mu_m) - D^{n-1} v_{m-1}(t-\mu_m)\right]. \tag{1.6}$$

In particular $(n = m)$, we obtain that

$$\max_{t\in(-a,a)} |D^m v_m(t)| \leqq \frac{1}{\mu_m} \max_{t\in(-a,a)} |D^{m-1} v_{m-1}(t)|.$$

Thus, the inequality (1.5) is valid if this inequality is valid for m-1 too. So the inequality (1.5) is valid for all m.

Let $n \leqq m-1$ be arbitrary now. In this case, using the inequality (1.6), we obtain that for any point $t\in(-a,a)$ there exist points $t_1,\ldots,t_{m-n}$ so that

$$D^n v_m(t) = D^n v_{m-1}(t_1) = \ldots = D^n v_n(t_{m-n}).$$

Consequently, from this and inequality (1.5) we have

$$\max_{t\in(-a,a)} |D^n v_m(t)| \leqq \max_{t\in(-a,a)} |D^n v_n(t)| \leqq (\mu_0\mu_1\ldots\mu_n)^{-1}. \tag{1.7}$$

From this, using Arzelà'stheorem and a diagonal process, we obtain that there exists a subsequence of sequence $v_m(t)$ (we shall denote it as $v_m(t)$ too) and a function $v(t) \in C_0^\infty(-a,a)$ so that

$$v_m(t) \longrightarrow v(t),\ldots, D^n v_m(t) \longrightarrow D^n v(t),\ldots$$

uniformly for $t\in(-a,a)$.

It is evident that v(t) is a function to be found. It only remains to remark that inequality (1.5) may be obtained from inequality (1.7) when $m \to \infty$. Lemma is proved.

Let us continue the proof of the theorem. Namely, using Lemma 1.1, we shall find a function $u(x) \in \overset{\circ}{W}{}^{\infty}\{a_\alpha, p_\alpha\}$. For this we shall first remark that according to the Mandelbrojt-Bang criterion of non-quasianalyticity the sequence $M_N$ satisfies the conditions

$$\lim_{N\to\infty} M_N^{\frac{1}{N}} = \infty , \quad \sum_{N=1}^{\infty} \frac{M_{N-1}^c}{M_N^c} < \infty ,$$

where $M_N^c$ is the convex regularization of sequence $M_N$ by means of logarithms (see S. MANDELBROJT [1]). Obviously, there exist natural number $N_0$, positive numbers $\mu_1, \dots, \mu_{N_0}$ and positive number $q < 1$ so that

$$\mu_1 + \dots + \mu_{N_0} + \sum_{N=N_0+1}^{\infty} \frac{1}{q} \frac{M_{N-1}^c}{M_N^c} < \frac{a}{3}, \tag{1.8}$$

where a number $a > 0$ will bee deseribed below more precisely.

Let us put

$$\mu_N = \frac{1}{q} \frac{M_{N-1}^c}{M_N^c}, \quad N = N_0+1, \dots$$

According to (1.8) and Lemma 1.1 we can find a function of one real variable $v(t) \in C_0^\infty(-a,a)$ which satisfies (1.4). Taking into account the definition of numbers $\mu_0, \mu_1, \dots,$ we shall have inequality

$$|D^N v(t)| \leqq K q^N M_N^c, \tag{1.9}$$

where $K > 0$ is a constant not depending on $N = 0,1,\dots$

Let us complete the proof of theorem. Namely, let us choose a point $x^0 = (x_1^0, \dots, x_n^0) \in G$ and a number $a > 0$ so that the cube $U = \{x : |x_j - x_j^0| < a,\ j = 1,\dots,n\}$ is included entirely in G. Let us put

$$u(x) = v(x_1 - x_1^0) \dots v(x_n - x_n^0), \quad x \in U,$$

and let us prove that $u(x) \in \overset{\circ}{W}{}^{\infty}\{a_\alpha, p_\alpha\}$. In fact, from (1.9) for $\alpha = (\alpha_1, \dots, \alpha_n)$, $|\alpha| = N$, we shall have the inequality

$$|D^\alpha u(x)| \leqq K q^N M_{\alpha_1}^c \dots M_{\alpha_n}^c , \quad K > 0.$$

Further, in view of the logarithmic convexity of sequence $M_N^c$, the inequality $M_{\alpha_1}^c \dots M_{\alpha_n}^c \leqq M_{\alpha_1+\dots+\alpha_n}^c = M_N^c$ (without loss of generality one can assume that $M_0^c = 1$) is valid. Consequently, our function u(x) satisfies the inequalities

$$|D^\alpha u(x)| \leqq Kq^N M_N^c, \quad |\alpha| = N, \quad x \in U,$$

where $K > 0$ is a constant, and a fortiori the inequalities

$$|D^\alpha u(x)| \leqq Kq^N M_N, \quad |\alpha| = N,$$

since $M_N^c \leqq M_N$.

From this, in view of equation (1.1), we find that

$$\varrho(u) \equiv \sum_{N=0}^{\infty}\left(\sum_{|\alpha|=N} a_\alpha \|D^\alpha u\|_{r_\alpha}^{p_\alpha}\right) \leqq \sum_{N=0}^{\infty}\left(\sum_{|\alpha|=N}^{\infty} a_\alpha K_1^{p_\alpha} q^{Np_\alpha} \cdot M_N^{p_\alpha}\right)$$

$$\leqq K_2 \sum_{N=0}^{\infty} q_0^N \left(\sum_{|\alpha|=N} a_\alpha M_N^{p_\alpha}\right) \equiv K_2 \sum_{N=0}^{\infty} q_0^N < \infty ,$$

where $q < q_0 < 1$; $K_1 > 0$, $K_2 > 0$ are some constants. Thus, $u(x) \in \mathring{W}^\infty\{a_\alpha, p_\alpha\}$, and the theorem is completely proved.

Example. Let $p_\alpha \equiv p$, $r_\alpha \equiv p$, where $p \geqq 1$ is a number. Then the sequence $M_N$, $N = 0,1,\dots,$ can be found as

$$M_N = \left(\sum_{|\alpha|=N} a_\alpha\right)^{-\frac{1}{p}}.$$

In particular, if $G = (a,b)$ is an interval, $a_N = (N!)^{-q}$, then the Banach space

$$\mathring{W}^\infty\left\{\frac{1}{(n!)^q}, p\right\} = \left\{u(x) \in C_0^\infty(a,b): \sum_{n=0}^{\infty} \frac{1}{(n!)^q} \|D^n u\|_p^p < \infty\right\}$$

is nontrivial precisely when $q > p$.

## § 2. Criterion for nontriviality of the spaces $W^\infty\{a_\alpha, p_\alpha\}$ in the case of full Euclidean space

Let $a_\alpha > 0$, $p_\alpha \geqq 1$, $r_\alpha \geqq 1$ be arbitrary sequences of real numbers, where indices $\alpha = (\alpha_1,\dots,\alpha_n)$, $\alpha_j \geqq 0$, $j = 1,\dots,n$, run through an unbounded set of values.

We shall consider the following space of functions $u(x): \mathbb{R}^n \to \mathbb{C}^1$

$$W^{\infty}\{a_\alpha, p_\alpha\} = \Big\{u(x):\ D^\alpha u \in L_{r_\alpha}(\mathbb{R}^n),$$

$$\varrho(u) \equiv \sum_\alpha a_\alpha \|D^\alpha u\|_{r_\alpha}^{p_\alpha} < \infty\Big\},$$

where $\|\cdot\|_r$ is the norm in $L_r(\mathbb{R}^n)$, $D^\alpha$ denotes the generalized derivative of order $\alpha$. In the case $r_\alpha = \infty$ we set $\|v\|_\infty = \operatorname{ess\,sup}|v(x)|$, $x \in \mathbb{R}^n$.

Definition 2.1. The space $W^{\infty}\{a_\alpha, p_\alpha\}$ is called nontrivial if there exists at least one function $u(x) \not\equiv 0$ so that $\varrho(u) < \infty$.

Our aim is to obtain necessary and sufficient conditions for nontriviality of the spaces $W^{\infty}\{a_\alpha, p_\alpha\}$. It is obvions that this problem becomes interesting if $a_0 > 0$, since otherwise the function $u(x) \equiv \text{const}$ belongs to $W^{\infty}\{a_\alpha, p_\alpha\}$. In view of this we shall assume that $a_0 > 0$.

Theorem 2.1. The space $W^{\infty}\{a_\alpha, p_\alpha\}$ is nontrivial if and only if there exists a point $q = (q_1, \ldots, q_n)$, $q_1 > 0, \ldots, q_n > 0$, so that

$$\sum_\alpha a_\alpha q^{\alpha p_\alpha} \equiv \sum_\alpha a_\alpha q_1^{\alpha_1 p_\alpha} \cdots q_n^{\alpha_n p_\alpha} < \infty . \qquad (*)$$

Remark. Before proving this theorem we shall note that in the case $p_\alpha \leqq \text{const}$ this theorem may be formulated as follows:

Corollary 2.1. Let $p_\alpha \geqq 1$ be a bounded sequence. Then the space $W^{\infty}\{a_\alpha, p_\alpha\}$ is nontrivial precisely when the function

$$\varphi(z) \equiv \sum_{N=0}^{\infty} b_N z^N \quad (b_N = \sum_{|\alpha|=N} a_\alpha)$$

is an analytic function at the point $z = 0$.

In other words, in the case of bounded values of $p_\alpha$ the space $W^{\infty}\{a_\alpha, p_\alpha\}$ is nontrivial if and only if

$$\overline{\lim_{N\to\infty}} \Big(\sum_{\alpha = N} a_\alpha\Big)^{\frac{1}{N}} < \infty .$$

Proof of theorem. Sufficiency. Let the condition (*) hold. We shall construct a function $u(x) \in W^{\infty}\{a_\alpha, p_\alpha\}$ which is not identical zero. In fact, let us consider a function $\tilde{u}(\xi) \in C_0^{\infty}(G)$ where

$$G = \left\{ \xi \in \mathbb{R}^n: \ |\xi| \leqq \xi_0 < q_0 \right\}, \quad q_0 = \min(q_1,\dots,q_n).$$

We shall show that

$$u(x) = \int_G \tilde{u}(\xi) e^{ix\xi} \, d\xi$$

is a function as desired.

In fact, after simple calculations we shall have the inequality

$$|D^\alpha u(x)| \leqq K\cdot \xi_0^{|\alpha|}\cdot |\alpha|^m \prod_{K=1}^{m} \min\left(1, \frac{1}{|x_K|^{m/n}}\right),$$

where multiindices $\alpha$ and $m \in \mathbb{N}$ are arbitrary; $K = K(m)$ is a constant, depending on $m$, $q_0$, $\max |D^\beta \tilde{u}(\xi)|$ for $|\beta| \leqq m$, but not depending on $\alpha$. Choosing $m = n+1$, we obtain that $D^\alpha u(x) \in L_{r_\alpha}(\mathbb{R}^n)$ and for any $\alpha$

$$\|D^\alpha u\|_{r_\alpha} \leqq K\cdot \xi_0^{|\alpha|} \cdot |\alpha|^m,$$

where $K > 0$ is a constant.

From the latter inequality and using the condition (*) we obtain that

$$\varrho(u) \equiv \sum_\alpha a_\alpha \|D^\alpha u\|_{r_\alpha}^{p_\alpha} \leqq \sum_\alpha (K\cdot \xi_0^{|\alpha|}\cdot |\alpha|^m)^{p_\alpha} < \infty,$$

as $\xi_0 < q_0$ (recall that $q_0 = \min(q_i,\dots,q_n)$, where $q_1 > 0,\dots,q_n > 0$ are from condition (*)).

This means that $u(x) \in W^\infty\{a_\alpha, p_\alpha\}$. Sufficiency of condition (*) is proved.

Necessity of condition (*). We shall show that if for any $q = (q_1,\dots,q_n)$, $q_1 > 0,\dots,q_n > 0$, the series (*) diverges, then every function $u(x) \in W^\infty\{a_\alpha, p_\alpha\}$ is identically zero.

The foundation of our proof is the following lemma which has (as we hope) an independent interest.

Lemma 2.1. Let the series (*) diverge for a fixed $q = (q_1,\dots,q_n)$, $q_1 > 0,\dots,q_n > 0$, that is

$$\sum_\alpha a_\alpha q^{\alpha p_\alpha} = +\infty.$$

Then for every function $u(x) \in W^\infty\{a_\alpha, p_\alpha\}$ the Fourier transform $\tilde{u}(\xi)$

is a usual generalized function. Moreover, $\tilde{u}(\xi) \equiv 0$ if $\xi \in G_q$, where $G_q = \left\{\xi : |\xi_j| > q_j, j = 1,\dots,n\right\}$.

Proof of lemma. It is obvious that if for given q the series (*) deverges then one of the series

$$(**) \sum_{r_\alpha \leqq 2} a_\alpha q^{\alpha p_\alpha}, \quad (***) \sum_{r_\alpha > 2} a_\alpha q^{\alpha p_\alpha}$$

must diverge too.

Let us consider first the case where the series (**) diverges. As is well known, if $u(x) \in L_s(\mathbb{R}^n)$, where $1 \leqq s \leqq 2$, then its Fourier transform $\tilde{u}(\xi)$ is the usual function; moreover, $\tilde{u}(\xi) \in L_{s'}(\mathbb{R}^n)$, $s' = s/(s-1)$, and

$$\|\tilde{u}(\xi)\|_{s'} \leqq \|u(x)\|_s \tag{2.1}$$

(see, for example A. ZYGMUND [1], E. STEIN and G. WEISS /1/).

Consequently, for all $\alpha$ such that $r_\alpha \leqq 2$ the Fourier transform

$$\widetilde{D^\alpha u(x)} = (i\xi)^\alpha \tilde{u}(\xi)$$

is the usual function; moreover, $\widetilde{D^\alpha u(x)} \in L_{r'_\alpha}(\mathbb{R}^n)$. Hence, it follows that $\tilde{u}(\xi)$ for $\xi_j \neq 0$, $j = 1,\dots,n$, is also the usual function (we do not exclude the case $r_0 > 2$, therefore, we cannot assert the regularity of $\tilde{u}(\xi)$ for $\xi_j = 0$, since $\tilde{u}(\xi)$ is, a priori, only a generalized function).

Taking into account the inequality (2.1), we obtain that

$$\sum_{r_\alpha \leqq 2} a_\alpha \|\xi^\alpha \tilde{u}(\xi)\|_{r'_\alpha}^{p_\alpha} \leqq \sum_{r_\alpha \leqq 2} a_\alpha \|D^\alpha u\|_{r_\alpha}^{p_\alpha} < \infty, \tag{2.2}$$

since $u(x) \in W^\infty\{a_\alpha, p_\alpha\}$.

On the other hand, evidently

$$\sum_{r_\alpha \leqq 2} a_\alpha \|\xi^\alpha \tilde{u}(\xi)\|_{r'_\alpha}^{p_\alpha} \geqq \sum_{r_\alpha \leqq 2} a_\alpha q^{\alpha p_\alpha} \beta_q^{p_\alpha}, \tag{2.3}$$

where

$$\beta_q = \inf_\alpha \left(\int_{G_q} |\tilde{u}(\xi)|^{r'_\alpha} d\xi\right)^{\frac{1}{r'_\alpha}}, \quad G_q = \left\{\xi : |\xi_j| \geqq q_j, 1 \leqq j \leqq n\right\}.$$

Since the series (**) diverges, inequalities (2.2) and (2.3) are consistent if and only if $\beta_q = 0$. As it is easy to see, the latter is possible only if $u(x) \equiv 0$ in $G_q$. Lemma is proved in the case $r_\alpha \leqq 2$.

Let us consider now the case of divergence of series (***), assuming in this connection that $r_0 > 2$ (the case $r_0 \leq 2$ is much simpler). In this case the Fourier transform $\tilde{u}(\xi)$ is the generalized function defined by

$$\langle \tilde{u}(\xi), \tilde{\varphi}(\xi) \rangle = (2\pi)^n \langle u(x), \varphi(x) \rangle , \qquad (2.4)$$

where $\varphi(x) \in C_0^\infty(\mathbb{R}^n)$, $\tilde{\varphi}(\xi)$ is the classical Fourier transform of function $\varphi(x)$ (see, for example, I. M. GELFAND and G. E. ŠILOV [1]).

Since $u(x) \in L_{r_0}(\mathbb{R}^n)$ and the functions with compact support are dense in $L_{r'_0}(\mathbb{R}^n)$, equality (2.4) holds for any functions $\varphi(x) \in L_{r'_0}(\mathbb{R}^n)$ (we emphasize that $r'_0 = r_0/(r_0-1) < 2$).

Consequently, equality (2.4) determines $\tilde{u}(\xi)$ as a functional on the space $V_{r_0}$ which is the image of $L_{r'_0}(\mathbb{R}^n)$ under the Fourier transform F: $\varphi \to \tilde{\varphi}$. The space $V_{r_0}$ does not coinside with $L_{r_0}(\mathbb{R}^n)$, but (it is obvious) contains the smooth functions of compact support.

Thus, the Fourier transform $\tilde{u}(\xi)$ of the function $u(x) \in L_{r_0}(\mathbb{R}^n)$ is the usual generalized function, that is the functional on $\mathcal{D}$ ($\mathcal{D}$ is the classical basic space).

Let us now show (ans it is very important) that $\tilde{u}(\xi) \equiv 0$ in the domain $G_q$.

Let G be an arbitrary domain with a smooth boundary $\Gamma$, which is contained in the parallelepiped $K_s^+ = \left\{ \xi : \xi_j \geq s_j > q_j, \ j = 1,\dots,n \right\}$. Choosing in (2.4)

$$\varphi(\xi) = \tilde{v}(\xi)\tilde{w}(\xi),$$

where $\tilde{v}(\xi) \in C_0^\infty(G)$ is a fixed function and $\tilde{w}(\xi) \in C_0^\infty(G)$ is an arbitrary function, we obtain the equality

$$\langle \tilde{v}(\xi)\tilde{u}(\xi), \tilde{w}(\xi) \rangle = (2\pi)^n \langle u(x), \varphi(x) \rangle , \qquad (2.5)$$

where $\varphi(x) = F^{-1}[\tilde{v}(\xi)w(\xi)] = v(x) * w(x) \in L_{r'_0}(\mathbb{R}^n)$.

The functional $\tilde{v}(\xi)\tilde{u}(\xi)$ has, obviously, a compact support. Thus, this functional as a generalized function has a finite order of singularity, i. e. there exist a number $m \in \mathbb{N}$ and ordinary functions $h_\alpha(\xi)$, $|\alpha| \leq m$, so that the function $\tilde{v}(\xi)\tilde{u}(\xi)$ can be represented in the form

$$\tilde{v}(\xi)\tilde{u}(\xi) = \sum_{|\alpha| \leq m} D^\alpha h_\alpha(\xi). \qquad (2.6)$$

Without loss of generality one can suppose that the functions $h_\alpha(\xi) \in C^1(G)$. Hence, it follows that there exists a function $\tilde{z}(\xi) \in C^m(G)$, $D^\alpha z|_\Gamma = 0$, $|\alpha| < m$, such that

$$\tilde{v}(\xi)\tilde{u}(\xi) = L_{2m}\tilde{z}(\xi) \equiv \sum_{|\alpha| \leq m} (-1)^{|\alpha|} D^{2\alpha}\tilde{z}(\xi)$$

(it is clear that the function $\tilde{z}(\xi)$ is the solution of the boundary value problem

$$L_{2m}\tilde{z}(\xi) = \sum_{|\alpha| \leq m} D^\alpha h_\alpha(\xi), \qquad \xi \in G,$$

$$D^\alpha \tilde{z}|_\Gamma = 0, \quad |\alpha| < m;$$

the advantage of the latter representation compared to (2.6) is that it is unique).

From (2.5) we now have the following equality

$$\langle \tilde{v}(\xi)\tilde{u}(\xi), \tilde{w}(\xi)\rangle = \langle L_{2m}\tilde{z}(\xi), \tilde{w}(\xi)\rangle = (2\pi)^n \langle u(x), \varphi(x)\rangle ,$$

or, equivalently,

$$\langle \tilde{z}(\xi), L_{2m}\tilde{w}(\xi)\rangle = (2\pi)^n \langle u(x), \varphi(x)\rangle , \tag{2.7}$$

where $\tilde{w}(\xi) \in C_0^\infty(G)$ is an arbitrary function.

It is clear that the left side of (2.7) admits the closure up to an arbitrary function $\tilde{w}(\xi) \in C^{2m}(G) \cap C_0^m(G)$. Therefore, the right side of (2.7) admits the closure up to an arbitrary function of the same type too. We should note in this connection that

$$\varphi(x) = v(x) * w(x) \in L_{r_0'}(\mathbb{R}^n).$$

The same arguments for the function $D^\alpha u(x)$ give us the equality

$$\langle \tilde{z}(\xi), L_{2m}(\xi^\alpha \tilde{w}(\xi))\rangle = (2\pi)^n (-i)^{|\alpha|} \langle D^\alpha u(x), \varphi(x)\rangle , \tag{2.8}$$

where $\tilde{w}(\xi)$ belongs to the space $C^{2m}(G) \cap C_0^m(G)$ too.

Let us take the function $\tilde{w}(\xi) \equiv \tilde{w}\ (\xi)$ so that

$$L_{2m}(\xi^\alpha \tilde{w}_\alpha(\xi)) = q^\alpha \tilde{z}(\xi), \quad \xi \in G.$$

Then from (2.8) we shall have the inequality

$$q^\alpha \langle \tilde{z}(\xi), \tilde{z}(\xi)\rangle \leq (2\pi)^n \|D^\alpha u\|_{r_\alpha} \cdot \|\varphi_\alpha\|_{r_\alpha'} , \tag{2.9}$$

where $\varphi_\alpha(x) = v(x) * w_\alpha(x)$.

Since supp $\widetilde{w}_\alpha(\xi) \subset G \subset K_s^+ \subset G_q$ and $q_j < s_j$, $j = 1,\dots,n$, then for large $\alpha$

$$(2\pi)^n \|\varphi_\alpha\|_{r'_\alpha} \leqq (2\pi)^n \|v\|_1 \cdot \|w_\alpha\|_{r'_\alpha} \leqq 1;$$

therefore, without loss of generality we can suppose that $(2\pi)^n \|\varphi_\alpha\|_{r'_\alpha} \leqq 1$ for all indices $\alpha$. In that case from (2.9) we have the following inequality

$$\sum_{r_\alpha > 2} a_\alpha q^{\alpha p_\alpha} \langle \widetilde{z}(\xi), \widetilde{z}(\xi) \rangle^{p_\alpha} \leqq \sum_{r_\alpha > 2} a_\alpha \|D^\alpha u\|_{r_\alpha}^{p_\alpha} < \infty ,$$

which by condition (***) reduces to the identity $\widetilde{z}(\xi) = 0$ in the domain G. Therefore, $\widetilde{u}(\xi) \equiv 0$ in G too.

Since the domain $G \subset K_s^+$ and parallelepiped $K_s^+ = \{\xi : \xi_j > s_j > q_j, 1 \leqq j \leqq n\}$ are arbitrary, then $\widetilde{u}(\xi) \equiv 0$ in the domain $G_q^+ = \{\xi : \xi_j > q_j, 1 \leqq j \leqq n\}$.

In the same way it can be proved that $\widetilde{u}(\xi) \equiv 0$ in the other open octants of $G_q$. Lemma 2.1 is completely proved.

Let us finish the proof of our theorem. Namely, let u(x) be a function from $W^\infty\{a_\alpha, p_\alpha\}$ and the series (*) divergent for some $q = (q_1,\dots,q_n)$, where $q_j > 0$, $1 \leqq j \leqq n$. Then in view of Lemma 2.1 the Fourier transform $\widetilde{u}(\xi)$ is concentrated on the hyperplanes $\xi_j = 0$, $1 \leqq j \leqq n$. We shall prove that in reality $\widetilde{u}(\xi) = 0$. Indeed, let $\widetilde{\nu}(\xi) \in C_0^\infty(\mathbb{R}^n)$ be a function the support of which has a non-empty intersection with the hyperplane $\xi_j = 0$ (j is fixed) and which, on the contrary, does not intersect other coordinate hyperplanes of codimension one or more. The function $\widetilde{v}(\xi) \equiv \widetilde{\nu}(\xi)\widetilde{u}(\xi)$ can be represented as a tensor product

$$\widetilde{v}(\xi) = \sum_{K=0}^{m} \widetilde{v}_K(\xi_1,\dots,\xi_{j-1},\dots,\xi_n) \otimes \delta^{(K)}(\xi_j), \tag{2.10}$$

where $\widetilde{v}_K$ are generalized functions and $m \in \mathbb{N}$ is a number.

Further, since $u(x) \in L_{r_0}(\mathbb{R}^n)$ we also have $v(x) = \nu(x) * u(x) \in L_{r_0}(\mathbb{R}^n)$ (the notations are clear).

On the other hand, using (2.10), we obtain

$$v(x) = \sum_{K=0}^{m} v_K(x_1,\dots,x_{K-1},x_K,\dots,x_n) \otimes (-ix_j)^K$$

so that the inclusion $v(x) \in L_{r_0}(\mathbb{R}^n)$ is possible if and only if $v_K = 0$, $K = 1,\dots,m$. It follows that $\widetilde{v}(\xi) \equiv 0$ and we have immediately that $\widetilde{u}(\xi) \equiv 0$ on the hyperplanes $\xi_j = 0$, $1 \leqq j \leqq n$, except their intersection. Thus, the function $\tilde{u}(\xi)$ is concentrated on coordinate hyperplanes of codimension two or more.

Repeating the above arguments, we obtain that the generalized function $\tilde{u}(\xi)$ is concentrated in the point $\xi = 0$. It is well known that in this case the function $\tilde{u}(\xi)$ is a finite linear combination of the $\delta$-function and its derivatives. Therefore, the function $u(x)$ is a polynomial and since $u(x) \in L_{r_0}(\mathbb{R}^n)$ we have $u(x) \equiv 0$. This proves the necessity of condition (*) and hence Theorem 2.1 too.

Example. Let $p \geq 1$ be a number and $p_\alpha \equiv p$, $r_\alpha \equiv p$. Suppose the numbers $a_\alpha > 0$ for some unbounded set of indices $\alpha$; in this connection $a_0 > 0$.

If (and only if)

$$\overline{\lim_{N\to\infty}} \left( \sum_{|\alpha|=N} a_\alpha \right)^{\frac{1}{N}} < \infty ,$$

then the Banach space $W^\infty\{a_\alpha, p\}(\mathbb{R}^n)$ with norm

$$\|u\|^p \equiv \sum_\alpha a_\alpha \|D^\alpha u\|_p^p$$

is nontrivial.

## § 3. Criterion for nontriviality of the spaces $W^\infty\{a_\alpha, p_\alpha\}$ in the case of the torus

Let, as before, $a_\alpha \geq 0$, $p_\alpha \geq 1$, $r_\alpha \geq 1$ be arbitrary sequences of real numbers. Let us denote by $T^n$ the n-dimensional torus. Consider the space

$$W^\infty\{a_\alpha, p_\alpha\} = \left\{ u(x) \in C^\infty(T^n) : \varrho(u) \equiv \sum_{|\alpha|=0}^{\infty} a_\alpha \|D^\alpha u\|_{r_\alpha}^{p_\alpha} < \infty \right\},$$

where $u(x)$, $x = (x_1, \dots, x_n)$, is a periodic function of period $2\pi$.

As usual, the question of the nontriviality of the space $W^\infty\{a_\alpha, p_\alpha\}$ arises, i. e. the question of existence of at least one infinitely differentiable periodic function $u(x)$ such that $\varrho(u) < \infty$.

We are interestet only in the spaces $W^\infty\{a_\alpha, p_\alpha\}$ which have infinite dimension, i. e. which contain an infinite set of linearly independent periodic functions. Such spaces are called nontrivial.

Theorem 3.1. The space $W^{\infty}\{a_\alpha, p_\alpha\}$ is nontrivial if and only if there exists a sequence of distinct multi-indices $q_\nu = (q_{1\nu},\dots,q_{n\nu})$, $\nu = 0,1,\dots$, such that

$$\sum_{|\alpha|=0}^{\infty} a_\alpha q_\nu^{\alpha p_\alpha} < \infty \quad . \tag{*}$$

Proof. Obviously, condition (*) is sufficient. In fact, in this case the periodic functions $\exp(iq_\nu,x) \in W^{\infty}\{a_\alpha, p_\alpha\}$, $\nu = 0,1,\dots$, since

$$\varrho(\exp(iq_\nu,x)) \equiv \sum_{|\alpha|=0}^{\infty} a_\alpha \| D^\alpha u\|_{r_\alpha}^{p_\alpha} = \sum_{|\alpha|=0}^{\infty} a_\alpha q_\nu^{\alpha p_\alpha} (2\pi)^{\frac{np_\alpha}{r_\alpha}} < \infty \quad .$$

(It is clear that the set of convergence of this series is the same as the set of convergence of the series (*).)

Proof of necessity. Let us assume the contrary: the series (*) converges for not more than a finite set of multi-indices $q = (q_1,\dots,q_n)$, namely, for $|q_1| < N_1,\dots,|q_n| < N_n$, where $N_j$ are integers. We profe in this case that

$$W^{\infty}\{a_\alpha, p_\alpha\} \subset L(\exp(iq,x)),$$

where $L(\exp(iq,x))$ is the linear hull of the functions $\exp(iq,x)$ with $|q_j| \leqq N_j$, $j = 1,\dots,n$. The latter contradicts the condition of the infinite dimensionality of $W^{\infty}\{a_\alpha, p_\alpha\}$.

In fact, if $u(x) \in W^{\infty}\{a_\alpha, p_\alpha\}$, $x \in T^n$, then $u(x)$ is an infinitely differentiable periodic function and, therefore,

$$u(x) = \sum_{|q|=0}^{\infty} C_q \exp(iq,x),$$

where

$$C_q = (2\pi)^{-n}(u, \exp(iq,x)).$$

Moreover, for any $\alpha$

$$D^\alpha u(x) = \sum_{|q|=0}^{\infty} C_q (iq)^\alpha \exp(iq,x),$$

and, consequently,

$$|C_q q^\alpha| = (2\pi)^{-n} |(D^\alpha u, \exp(iq,x))| \leqq (2\pi)^{\frac{n}{r'_\alpha}} \| D^\alpha u\|_{r_\alpha},$$

whence

$$\|D^\alpha u\|_{r_\alpha}^{p_\alpha} \geq |C_q\, q^\alpha|^{p_\alpha}. \qquad (3.1)$$

Further, our assumption means, in particular, that the series (*) diverges in the points of type $(0,\dots,N_j,\dots,0)$, $1 \leq j \leq n$.

It is clear that any multi-index complementary to the multi-indices q, where $|q_j| \leq N_j$, majorizes an index of this type, that is

$$(|q_1|,\dots,|q_j|,\dots,|q_n|) \geq (0,\dots,N_j+1,\dots,0),$$

where $j = 1,\dots,n$ is some integer.

It is obvious, that for any j

$$\varrho(u) \equiv \sum_{|\alpha|=0}^{\infty} a_\alpha \|D^\alpha u\|_{r_\alpha}^{p_\alpha} \geq \sum_{\alpha_j=0}^{\infty} a_\alpha \|D^\alpha u\|_{r_\alpha}^{p_\alpha},$$

where the indices $\alpha$ run through the set of values $(0,\dots,\alpha_j,\dots,0)$.

Hence, for every index q complementary to the indices q, where $|q_j| \leq N_j$, $1 \leq j \leq n$, we obtain the following inequality

$$\varrho(u) \geq \sum_{\alpha_j=0}^{\infty} a_\alpha |C_q\, q_j^{\alpha_j}|^{p_\alpha} \geq \sum_{\alpha_j \gg 1} a_\alpha N_j^{\alpha p_\alpha} = \infty$$

if only $C_q \neq 0$ (here we take into account inequality (3.1)).

Since the latter inequality is impossible, we obtain that for any $u(x) \in W^\infty\{a_\alpha, p_\alpha\}$ the Fourier coefficients $C_q$ equal zero, provided at least one $q_j > N_j$, $1 \leq j \leq n$.

In the end we obtain that, if the series (*) converges for only a finite set of the points $q = (q_1,\dots,q_n)$, the space $W^\infty\{a_\alpha, p_\alpha\}$ of periodic functions has finite dimension, contrary to our condition. The theorem is proved.

Remark 3.1. Condition (*) does not exclude the case when the space $W^\infty\{a_\alpha, p_\alpha\}$ contains only functions of n-1, n-2,... arguments. In order to exclude this degenerate case one may require that the space $W^\infty\{a_\alpha, p_\alpha\}$ be dense in $L_2(T^n)$. It is easy to see that this requirement is necessary and sufficient in order that condition (*) be satisfied for some sequence $q_\nu$ such that

$$\min(q_{1\nu},\dots,q_{n\nu}) \to \infty$$

as $\nu \to \infty$. In the future we consider only such spaces.

Corollary 3.1. If the sequence $p_\alpha \geqq 1$ is bounded, then the space $W^\infty\{a_\alpha, p_\alpha\}$, $x \in T^n$, is nontrivial precisely when the function

$$\varphi(z) \equiv \sum_{N=0}^{\infty} b_N z^N \qquad (b_N = \sum_{|\alpha|=N} a_\alpha)$$

is an entire function of the complex argument z.

In particular, for $p_\alpha \equiv r_\alpha \equiv p$ this gives a criterion for the nontriviality of Banach space $W^\infty\{a_\alpha, p\}$, $x \in T^n$, with norm

$$\|u\|^p = \sum_{|\alpha|=0}^{\infty} a_\alpha \|D^\alpha u\|_p^p .$$

## § 4. Criterion for nontriviality of the spaces $\overset{\circ}{W}{}^\infty\{a_{n\alpha}, p\}$ in the case of a strip

Let $t \in [0,a]$, $x \in \mathbb{R}^\nu$ and $G = [0,a] \times \mathbb{R}^\nu$ be a strip in the space $\mathbb{R}^{\nu+1}$. We consider the space

$$\overset{\circ}{W}{}^\infty\{a_{n\alpha}, p\} = \Big\{ u(t,x) \in C^\infty(G) : \varrho(u)$$

$$\equiv \sum_{n+|\alpha|=0}^{\infty} a_{n\alpha} \|D_t^n D_x^\alpha u\|_p^p < \infty \ ; \ D_t^m u(0,x) = 0,$$

$$D_t^m u(a,x) = 0, \quad m = 0,1,\ldots \Big\},$$

where $a_{n\alpha} > 0$ is an arbitrary sequence of numbers and the number $p \geqq 1$. We can assume without loss of generality that the coefficient $a_{00} > 0$ (cf. § 2).

Definition 4.1. A space $\overset{\circ}{W}{}^\infty\{a_{n\alpha}, p\}$ is said to be nontrivial (nonempty) if it contains at least one function u(t,x) not identically equal to zero.

Theorem 4.1. A space $\overset{\circ}{W}{}^\infty\{a_{n\alpha}, p\}$ is nontrivial if and only if the following conditions are fulfilled:

a) There exists a point $q = (q_1, \ldots, q_\nu)$, $q_1 > 0, \ldots, q_\nu > 0$, such that

$$b_n = \sum_{|\alpha|=0}^{\infty} a_{n\alpha} q^{\alpha p} < \infty \quad , \ n = 0,1,\ldots;$$

b) The sequence $M_n = \left\{ b_n^{-\frac{1}{p}} \text{ if } b_n > 0;\ +\infty,\ \text{if } b_n = 0 \right\}$ defines a non-quasianalytic Hadamard class of one real variable (see § 1).

Proof. Let us suppose that conditions a) and b) are fulfilled. We will find a function u(t,x) from the space $\overset{\circ}{W}{}^{\infty}\{a_{n\alpha},p\}$ that is not identically equal to zero.

For this we consider the ball $S = \{\xi \in \mathbb{R}^{\nu},\ |\xi| \leq \xi_0 < Q_0$, where $q_0 = \min(q_1,\dots,q_{\nu})$, and put

$$v(x) = \int_S \tilde{v}(\xi) e^{ix\xi} d\xi,$$

where $\tilde{v}(\xi) \in C_0^{\infty}(S)$.

In conformity with the condition b) there exists a nontrivial function $\varphi(t) \in C_0^{\infty}(0,a)$ such that

$$|D_t^n \varphi(t)| \leq Kp^n M_n^c,\quad t \in (0,a), \tag{4.1}$$

where K > 0 and $q \in (0,1)$ are constants, $M_n^c$ is a convex regularization of the sequence $M_n$ by means of logarithms (see Lemma 1.1 in § 1).

Let us show that $u(t,x) = \varphi(t)v(x)$ is the desired function. Indeed, a simple calculation gives the following inequality for any natural number m and any :

$$|D_x^{\alpha} v(x)| \leq K \xi_0^{|\alpha|} |\alpha|^m \prod_{K=1}^{\nu} \min(1, |x_K|^{-\frac{m}{\nu}}),\quad x \in \mathbb{R}^{\nu},$$

where K > 0 depends in m, $q_0$ and $\max |D_{\xi}^{\beta} \tilde{v}(\xi)|$ when $|\beta| \leq m$, but not on $\alpha$. Putting $m = \nu+1$, we get that $D_x^{\alpha} v(x) \in L_p(\mathbb{R}^{\nu})$ and

$$\| D_x^{\alpha} v \|_{L_p(\mathbb{R}^{\nu})} \leq K \xi_0^{|\alpha|} |\alpha|^m$$

for all $\alpha$, where K > 0 is a constant.

Using the above estimate and taking into account the inequality (4.1), we immediately deduce that

$$\rho(u) \equiv \sum_{n+|\alpha|=0}^{\infty} a_{n\alpha} \| D_t^n D_x^{\alpha} u \|_p^p$$

$$\equiv \sum_{n+|\alpha|=0}^{\infty} a_{n\alpha} \| D_t^n \varphi(t) \|_{L_p(0,a)}^p \| D_x^{\alpha} v(x) \|_{L_p(\mathbb{R}^{\nu})}^p$$

$$\leqq K \sum_{n=0}^{\infty} q^{np} (M_n^c)^p \sum_{|\alpha|=0}^{\infty} a_{n\alpha} \xi_0^{|\alpha|\cdot p} |\alpha|^{mp}$$

$$\leqq K \sum_{n=0}^{\infty} q^{np} (M_n^c)^p b_n \leqq K \sum_{n=0}^{\infty} q^{np} < \infty .$$

It follows that $u(t,x) \in \mathring{W}^{\infty}\{a_{n\alpha}, p\}$. The sufficiency is proved.

<u>Necessity.</u> Suppose that either condition a) or condition b) is not satisfied. The necessity will be proved if we show that any function $u(t,x) \in \mathring{W}^{\infty}\{a_{n\alpha}, p\}$ is identically equal to zero.

Assume first that condition a) is not satisfied, i. e. for any $q = (q_1, \ldots, q_\nu)$, $q_1 > 0, \ldots, q_\nu > 0$, there exists a number $n$ such that

$$\sum_{|\alpha|=0}^{\infty} a_{n\alpha} q^{\alpha p} = +\infty . \tag{4.2}$$

The inclusion $u(t,x) \in \mathring{W}^{\infty}\{a_{n\alpha}, p\}$ implies, in particular, that for this $n$

$$\sum_{|\alpha|=0}^{\infty} a_{n\alpha} |D_x^{\alpha}(D_t^n u)|_p^p < \infty .$$

From this it follows that

$$\sum_{|\alpha|=0} a_{n\alpha} \|D_x^{\alpha}(D_t^n u)\|_{L_p(\mathbb{R}^\nu)} < \infty$$

for almost every $t \in (0,a)$, i. e. $D_t^n u \in W^{\infty}\{a_{n\alpha}, p\}(\mathbb{R}^\nu)$ for almost every $t \in (0,a)$. In conformity with the criterion for nontriviality of the spaces $W^{\infty}\{a_{n\alpha}, p\}(\mathbb{R}^\nu)$ (see § 2, chapter I), condition (4.2) implies that $D_t^n u(t,x) \equiv 0$ for almost every $t \in (0,a)$. Since $u(t,x)$ is an infinitely differentiable function and $D_t^m u(0,x) = 0$, $m = 0,1,\ldots$ then $u(t,x) \equiv 0$ with respect to $t$. Q.E.D.

Suppose now that condition b) is not satisfied. This means that there exists a point $q = (q_1, \ldots, q_\nu)$, $q_1 > 0, \ldots, q_\nu > 0$, such that all of the $b_n$ are finite but the sequence

$$M_n = \left\{ b_n^{-\frac{1}{p}} \text{ if } b_n > 0; \ +\infty \text{ if } b_n = 0 \right\}$$

defines a quasianalytic Hadamard class. We want to show that in this case the inclusion $u(t,x) \in \mathring{W}^{\infty}\{a_{n\alpha}, p\}$ implies that $u(t,x) \equiv 0$ too.

It is useful to consider the two cases $p \leqq 2$ and $p > 2$.

I (the case p = 2). Let $D_x^\alpha \tilde{u}(\tau,x)$ be the Fourier transform of $D_x^\alpha u(t,x)$ with respect to the variable t. In view of the fact that u(t,x) is a finite function with respect to t, we get that the following inequality holds for any $n \geq 2$ and for all $\tau \in \mathbb{R}^1$, $x \in \mathbb{R}^\nu$

$$|\tau|^{n-2}(1 + \tau^2)|D_x^\alpha u(\tau,x)| \leq K\|D_t^n D_x^\alpha u\|_{L_p(0,a)}, \tag{4.3}$$

where K > 0 is a constant.

Consequently, $D_x^\alpha \tilde{u}(\tau,x) \in L_p(\mathbb{R}^\nu)$ for any $\tau \in \mathbb{R}^1$ and

$$a_{n\alpha}|\tau|^{(n-2)p}(1 + \tau^2)\|D_x^\alpha \tilde{u}(\tau,x)\|_{L_p(\mathbb{R}^\nu)} = K\varrho(u). \tag{4.4}$$

As noted in § 2, for any function $u(x) \in L_p(\mathbb{R}^\nu)$, $p \leq 2$, its Fourier transform $\tilde{u}(\xi)$ is a usual function belonging to $L_{p'}(\mathbb{R}^\nu)$, $p' = p/(p-1)$, moreover

$$\|\tilde{u}(\xi)\|_{L_{p'}(\mathbb{R}^\nu)} \leq \|u(x)\|_{L_p(\mathbb{R}^\nu)}.$$

Consequently, going over now to the Fourier transform with respect to x, we have from (4.3) the following inequality

$$|\tau|^{p(n-2)}(1 + \tau^2)^p \sum_{|\alpha|=0}^{\infty} a_{n\alpha}\|(i\xi)^\alpha \tilde{u}(\tau,\xi)\|^p_{L_{p'}(\mathbb{R}^\nu)} < \infty . \tag{4.5}$$

Let us consider the domain $G_q = \{\xi : |\xi_j| \geq q_j,\ j = 1,\dots,\nu$, where $q = (q_1,\dots,q_\nu)$ is a vector for which the numbers bn define a quasianalytic class $C(M_n)$ (see condition a) of our theorem). It is clear that from the inequality (4.5) we have

$$b_n(1 + \tau^2)^p|\tau|^{(n-2)p}\|\tilde{u}(\tau,\xi)\|^p_{L_{p'}(G_q)} \leq K\varrho(u),$$

and, consequently,

$$|\tau|^{n-2}\|\tilde{u}(\tau,\xi)\|_{L_{p'}(G_q)} \leq \frac{K\ M_n}{1 + \tau^2},\ \tau \in \mathbb{R}^1,$$

where K > 0 is a constant and $n \geq 2$ is arbitrary.

From this inequality we obtain, in particular, that the inverse Fourier transform v(t) of function $\tilde{v}(\tau) = \|\tilde{u}(\tau,\xi)\|_{L_{p'}(G_q)}$ satisfies the inequality

$$|D_t^n v(t)| \leq KM_{n+2},$$

where K > 0 is a constant and t is arbitrary. It follows that $v(t) \in C(M_{n+2})$, i. e. v(t) belongs to a Hadamard class.

On the other hand, since $u(t,x) \in L_p$ in the strip $G = (0,a) \times \mathbb{R}^\nu$ and is a finite function with respect t, it follows from the Paley-Wiener theorem that $v(t)$ is also a finite function with respect to t. Consequently, by assumption, $v(t) \equiv 0$, and hence $\tilde{u}(\tau,\xi) = 0$ for all $\tau \in \mathbb{R}^1$ and $\xi \in G_q$. Evidently, however, if condition b) is not true for some q, then it is also not true for all $q^* = (q_1^*,\dots,q_\nu^*)$ such that $0 < q_j^* \leq q_j$, $1 \leq j \leq \nu$. Thus $\tilde{u}(\tau,\xi) = 0$ almost everywhere and hence $u(t,x) = 0$ everywhere too. Q.E.D.

II (the case $p > 2$). In this case we want to use the Fourier transform also. However, in this case the Fourier transform of function $u(x) \in L_p(\mathbb{R}^\nu)$ is (in general) the distribution $\tilde{u}(\xi)$ defined by

$$\langle \tilde{u}(\xi), \tilde{\varphi}(\xi) \rangle = (2\pi)^\nu \langle u(x), \varphi(x) \rangle , \tag{4.6}$$

where $\varphi(x) \in C_0^\infty(\mathbb{R}^\nu)$ and $\tilde{\varphi}(\xi)$ is its classical Fourier transform.

Lemma 4.1 (cf. with lemma 2.1 in § 2 of the present Chapter). Let $u(t,x) \in \overset{\circ}{W}{}^\infty\{a_{n\alpha}, p\}$ be an arbitrary function. Then for all $\tau \in \mathbb{R}^1$ the Fourier transform $\tilde{u}(\tau,\xi)$ of $u(t,x)$ is a usual distribution (i. e. a distribution on the space D of test functions) with support contained in the coordinate hyperplanes $\xi_j = 0$, $j = 1,\dots,\nu$.

Proof. Indeed, in view of the inclusion $u(t,x) \in L_p(G)$, where $G = (0,a) \times \mathbb{R}^\nu$ is our strip, we have that for all $\tau$

$$\langle \tilde{u}(\tau,\xi), \tilde{\varphi}(\xi) \rangle = (2\pi)^\nu \int_0^a \langle u(t,x), \varphi(x) \rangle \, e^{it\tau} \, dt, \tag{4.7}$$

where $\varphi(x) \in C_0^\infty(\mathbb{R}^\nu)$ is an arbitrary function.

As is known, the finite functions are dense in $L_{p'}(\mathbb{R}^\nu)$ and, consequently, the relation (4.7) can be extended to any function $\varphi(x) \in L_{p'}(\mathbb{R}^\nu)$. Consequently, the left side of (4.7) defines $\tilde{u}(\tau,\xi)$ as a functional on the image $V_p$ of the space $L_{p'}(\mathbb{R}^\nu)$ under the Fourier transform $F: \varphi(x) \longrightarrow \tilde{\varphi}(\xi)$ (we remark that $p' < 2$ and, consequently, $\tilde{\varphi}(\xi)$ is a usual function). The space $V_p$ contains arbitrary finite functions, and hence $\tilde{u}(\tau,\xi)$, as a function of $\xi$, is a distribution on D.

Let us now prove the main assertion of the lemma.

Let the numbers $q_j > 0$, $j = 1,\dots,\nu$, be defined by condition b) of our theorem and the $s_j \geq q_j$ arbitrary. Let $\Omega$ be an arbitrary domain with smooth boundary $\Gamma$ that is contained in the parallelepiped

$$\Pi = \left\{ \xi : q_j \le \xi_j \le s_j,\ 1 \le j \le \nu \right\}.$$

Putting in (4.7) $\widetilde{\varphi}(\xi) = \widetilde{v}(\xi)\widetilde{w}(\xi)$, where $\widetilde{v}(\xi) \in C_0^\infty(\Omega)$ is a fixed function and $\widetilde{w}(\xi) \in C_0^\infty(\Omega)$ is an arbitrary function, we obtain the relation

$$\langle \widetilde{v}(\xi)\widetilde{u}(\tau,\xi), \widetilde{w}(\xi) \rangle = (2\pi)^\nu \int_0^a \langle u(t,x), \varphi(x) \rangle\, e^{it\tau}\, d\tau, \quad (4.8)$$

where

$$\varphi(x) = F^{-1}\left[\widetilde{v}(\xi)\widetilde{w}(\xi)\right] = v(x) * w(x) \in L_{p'}(\mathbb{R}^\nu).$$

The distribution $\widetilde{v}(\xi)\widetilde{u}(\tau,\xi)$ has a compact support and, therefore, it has singularities of finite order. It follows that

$$\widetilde{v}(\xi)\widetilde{u}(\tau,\xi) = \sum_{|\beta| \le m} D^\beta h_\beta(\tau,\xi), \quad (4.9)$$

where $h_\beta(\tau,\xi)$ are usual functions in $\Omega$ (e. g. $h_\beta(\tau,\xi) \in C^1(\Omega)$) and m is a nonnegative number. We note that m can be chosen independent of $\tau$, since $u(t,x)$ is a smooth function with respect to $t \in (0,a)$.

Let $\widetilde{z}(\tau,\xi) \in C^m(\Omega)$, $D_\xi^\beta \widetilde{z}(\tau,\xi)\big|_\Gamma = 0$, $|\beta| \le m-1$, be a function such that

$$\widetilde{v}(\xi)\widetilde{u}(\tau,\xi) = L_{2m}\, \widetilde{z}(\tau,\xi) = \sum_{|\beta| \le m} (-1)^{|\beta|}\, D^\beta (D^\beta \widetilde{z}(\tau,\xi))$$

($\widetilde{z}(\tau,\xi)$ is a solution of the Dirichlet problem

$$L_{2m}\, \widetilde{z}(\tau,\xi) = \sum_{|\beta| \le m} D^\beta h_\beta(\tau,\xi)$$

$$D^\beta \widetilde{z}(\tau,\xi)\big|_\Gamma = 0, \quad |\beta| \le m-1;$$

the advantage of the latter representation compared with (4.9) is that it is unique).

From (4.8) we now have

$$\langle \widetilde{v}(\xi)\widetilde{u}(\tau,\xi), \widetilde{w}(\xi) \rangle = \langle L_{2m}\, \widetilde{z}(\tau,\xi), \widetilde{w}(\xi) \rangle =$$

$$= (2\pi)^\nu \int_0^a \langle u(t,x), \varphi(x) \rangle\, e^{it\tau}\, dt$$

or, equivalently,

$$\langle \widetilde{z}(\tau,\xi), L_{2m}\, \widetilde{w}(\xi) \rangle = (2\pi)^\nu \int_0^a \langle u(t,x), \varphi(x) \rangle\, e^{it\tau}\, dt, \quad (4.10)$$

where, we recall, $\widetilde{w}(\xi) \in C_0^\infty(\Omega)$ is an arbitrary function. It is clear that the left side of (4.10) admits the closure up to an arbitrary function $\widetilde{w}(\xi) \in C^{2m}(\Omega) \cap C_0^m(\Omega)$. Therefore, the right side of (4.10) admits the closure up to an arbitrary function of the same type too. We should note in this connection that

$$\varphi(x) = v(x)*w(x) \in L_{p'}(\mathbb{R}^\nu).$$

The same arguments for the function $D_t^n D_x^\alpha u$ lead to the relation

$$\tau^n \langle \widetilde{z}(\tau,\xi), L_{2m}\xi^\alpha \widetilde{w}(\xi)\rangle$$

$$= (2\pi)^\nu (-i)^{|\alpha|+n} \int_0^a \langle D_t^n D_x^\alpha u(t,x), \varphi(x)\rangle e^{it\tau} dt, \tag{4.11}$$

where $\widetilde{w}(\xi) \in C^{2m}(\Omega) \cap C_0^m(\Omega)$.

Let us take the function $\widetilde{w}(\xi) \equiv \widetilde{w}_\alpha(\tau,\xi)$ so that

$$L_{2m}(\xi^\alpha \widetilde{w}_\alpha(\tau,\xi)) = q_0^\alpha \widetilde{z}(\tau,\xi),$$

where $q_0 = (q_{01},\ldots,{}_{0\nu})$, $0 < q_{0j} < q_j$, $j = 1,\ldots,\nu$. Then from (4.11) we shall have the inequality

$$|\tau^n q_0^\alpha \langle \widetilde{z}(\tau,\xi), \widetilde{z}(\tau,\xi)\rangle|$$

$$\leqq K \|D_t^n D_x^\alpha u(t,x)\|_p \cdot \|\varphi_\alpha(\tau,x)\|_{L_p(\mathbb{R}^\nu)}, \tag{4.12}$$

where K is a positive constant and $\varphi_\alpha(\tau,x) = v(x)*w_\alpha(\tau,x)$.

It is easy to see that for large $|\alpha|$ and any $\tau$

$$\|\varphi_\alpha(\tau,x)\|_{L_{p'}(\mathbb{R}^\nu)} \leqq \|v(x)\|_{L_1(\mathbb{R}^\nu)} \cdot \|w_\alpha(\tau,x)\|_{L_{p'}(\mathbb{R}^\nu)} \leqq 1;$$

consequently, from (4.12) we get that

$$|\tau^n| q_0^\alpha \langle \widetilde{z}(\tau,\xi), \widetilde{z}(\tau,\xi)\rangle \leqq K \|D_t^n D_x^\alpha u\|_p.$$

Since u(t,x) is a test function with respect to t, we immediately obtain from this result the inequality

$$(1+\tau^2)|\tau|^{n-2} q_0^\alpha \langle \widetilde{z}(\tau,\xi), \widetilde{z}(\tau,\xi)\rangle \leqq K \|D_t^n D_x^\alpha u\|_p$$

for all $n \geqq 2$, and, as a consequence,

$$(1+\tau^2)|\tau|^{(n-2)p} b_n \langle \widetilde{z}(\tau,\xi), \widetilde{z}(\tau,\xi)\rangle^p \leqq K\varrho(u),$$

where the numbers $b_n$ are defined in the conditions of the theorem.

From this it follows that

$$|\tau|^n \langle \tilde{z}(\tau,\xi),\tilde{z}(\tau,\xi)\rangle \leq \frac{K}{1+\tau^2} b_n^{-\frac{1}{p}} \equiv \frac{K}{1+\tau^2} M_n,$$

where $K > 0$ is a constant. This means that the function

$$\psi(t) \equiv F^{-1}_{\tau\to t} \langle \tilde{z}(\tau,\xi),\tilde{z}(\tau,\xi)\rangle$$

satisfies the inequalities

$$|D_t^n \psi(t)| \leq K M_{n+2}, \quad n = 0,1,\ldots,$$

i.e. $\psi(t) \in C(M_{n+2})$, where $C(M_n)$ is a Hadamard class.

On the other hand, $\psi(t)$ is a finite function in view of the Paley-Wiener theorem. Since, by assumption, the class $C(M_{n+2})$ (together with the class $C(M_n)$) is quasianalytic, we have $\tilde{z}(\tau,\xi) \equiv 0$ for all $\xi \in \Omega$. Since the parallelepiped $\Pi$ and the function $\tilde{v}(\xi) \in C_0^\infty(\Omega)$ are arbitrary, we have $\tilde{z}(\tau,\xi) \equiv 0$ (and, consequently, $\tilde{u}(\tau,\xi) \equiv 0$) for all $\xi = (\xi_1,\ldots,\xi_\nu)$, $\xi_j > 0$, $j = 1,\ldots,\nu$.

By analogy one can prove that $\tilde{u}(\tau,\xi) \equiv 0$ interior to the other octants of $\mathbb{R}^\nu$. The lemma is proved.

We can now complete the proof of the theorem. It is convenient to put $\tau = \xi_0$ and regard $\tilde{u}(\tau,\xi)$ as the function $\tilde{u}(\xi_0,\xi)$ in $\mathbb{R}^{\nu+1}$.

From the lemma we obtain that the support of this distribution is contained in the coordinate hyperplanes $\xi_j = 0$, $j = 0,\ldots,\nu$. Let $G_j$ $(0 \leq j \leq \nu)$ be an arbitrary domain which has nonempty intersection with the hyperplane $\xi_j = 0$ and which, in contrast, does not intersect other coordinate hyperplanes of codimension one or more.

Consider the function $\tilde{v}(\xi_0,\xi) = \tilde{\nu}(\xi_0,\xi)\tilde{u}(\xi_0,\xi)$, where $\tilde{\nu}(\xi_0,\xi) \in C_0^\infty(G_j)$. In view of the lemma, the support of $\tilde{v}(\xi_0,\xi)$ is contained in the domain $V_j = G_j \cap \{\xi_j = 0\}$. Consequently, $\tilde{v}(\xi_0,\xi)$ is representable in the form

$$\tilde{v}(\xi_0,\xi) = \sum_{K=0}^{m} \tilde{v}_K(\xi_0,\ldots,\xi_{j-1},\xi_{j+1},\ldots,\xi_\nu) \otimes \delta^{(K)}(\xi_j), \quad (4.13)$$

where $\tilde{v}_K \in D'(V_j)$ and $m \geq 0$ is an integer.

Further, since $u(t,x) \in L_p(\mathbb{R}^{\nu+1})$, then $v(t,x) = \nu(t,x) * u(t,x) \in L_p(\mathbb{R}^{\nu+1})$, too.

On the other hand, the formula (4.13) gives

$$v(t,x) = \sum_{K=0}^{m} v_K(t,x_1,\ldots,x_{j-1},x_{j+1},\ldots,x_\nu)\cdot(-ix_j)^K,$$

so that the inclusion $v(t,x) \in L_p(\mathbb{R}^{\nu+1})$ is possible if and only if $v_K = 0$, $K = 0,\ldots,m$. This means that $\tilde{v}(\xi_0,\xi) = 0$ in $V_j$, $j = 0,\ldots,\nu$. In conformity with the choice of $V_j$ we obtain that $\tilde{u}(\xi_0,\xi) \equiv 0$ on the hyperplane $\xi_j = 0$, $0 \leq j \leq \nu$, with the possible exception of their intersection. Thus the support of $\tilde{u}(\xi_0,\xi)$ is concentrated on the hyperplanes of codimension two and more.

Repeating the above considerations, we get that the support of $\tilde{u}(\xi_0,\xi)$ is the point $\xi_0 = \xi_1 = \ldots = \xi_\nu = 0$ and, consequently, that $u(t,x)$ is a polynomial. Since $u(t,x) \in L_p(\mathbb{R}^{\nu+1})$, then $u(t,x) \equiv 0$.

The necessity of conditions a) and b) in the case when $p > 2$ is proved. The theorem is proved.

Remark. The above proof can also be carried out with elementary variations for the case of the space $\mathring{W}^\infty\{a_{n\alpha},p_n\}(G)$ when the numbers $p_n \geq 1$ are arbitrary.

Example. Consider the space $\mathring{W}^\infty\{a_{n\alpha},p\}(G)$, where $a_{n\alpha} = a_n c_\alpha$ ($a_n \geq 0$, $c_\alpha \geq 0$). In conformity with Theorem 4.1 this space is nontrivial if and only if the following two conditions are fulfilled:

a) There exists a point $q = (q_1,\ldots,q_\nu)$, $q_j > 0$, such that

$$\sum_{|\alpha|=0}^{\infty} c_\alpha q^{\alpha p} < \infty \ ;$$

b) The numbers

$$M_n = \left\{ a_n^{-\frac{1}{p}} \text{ if } a_n > 0;\ +\infty \text{ if } a_n = 0 \right\}$$

define a non-quasianalytic Hadamard class $C(M_n)$.

Taking into account the results of § 1 and § 2 of the present Chapter, we can reformulate our assertion in this case as follows: The space $\mathring{W}^\infty\{a_n c_\alpha,p\}(G)$, where $G = (0,a)\times\mathbb{R}^\nu$, is nontrivial if and only if the spaces $\mathring{W}^\infty\{a_n,p\}(0,a)$ and $W^\infty\{c_\alpha,p\}(\mathbb{R}^\nu)'$ are nontrivial.

## § 5. Further investigations

As it follows from §§ 1 - 3 the criteria of nontriviality of Sobolev spaces of infinite order depend essentially not only on the direct parameters $a_\alpha \geqq 0$, $p_\alpha \geqq 1$, $r_\alpha \geqq 1$, but on the region $G \subset \mathbb{R}^n$, where the functions $u(x) \in W^\infty\{a_\alpha, p_\alpha\}$ are defined, boundary conditions etc. Everything points to the fact that the question of the nontriviality of such spaces must be solved concretely in any case. Let us describe the further results in this direction which are known at present.

CHAN DYK VAN [4], [5] studied Sobolev-Orlicz classes of infinite order

$$L\overset{\circ}{W}{}^\infty = \left\{ u(x) \in C_0^\infty(G) : \sum_{|\alpha|=0}^{\infty} \int_G \varrho_\alpha (D^\alpha u)\, dx < \infty \right\},$$

where $\varrho_\alpha(\xi) \geqq 0$ are any N-functions (in the case $\varrho_\alpha(\xi) = a_\alpha |\xi|^{p_\alpha}$ we have the space $\overset{\circ}{W}{}^\infty\{a_\alpha, p_\alpha\}$). He obtained an elegant criterion for nontriviality of the classes $L\overset{\circ}{W}{}^\infty$. Namely, let

$$M_N = \inf_{|\alpha|=N} \varrho_\alpha^{-1}\left(\frac{1}{\text{mes } G}\right);$$

moreover, $M_N = +\infty$ if $\varrho_\alpha(\xi) \equiv 0$ for all $\alpha$ such as $|\alpha| = N$.

Then the class $L\overset{\circ}{W}{}^\infty$ is nontrivial if and only if the class of Hadamard $C\{M_N\}$ is non-quasianalytic. Using the known theorem of P. LELONG (see P. LELONG [1]) we may formulate this criterion in the following equivalent form: the class $L\overset{\circ}{W}{}^\infty$ is nontrivial if and only if the sequence.

$$M_\alpha = \varrho_\alpha^{-1}\left(\frac{1}{\text{mes } G}\right), |\alpha| = 0,1,\ldots$$

defines the non-quasianalytic Hadamard class of a function $u(x)$ of $n$ arguments $x = (x_1,\ldots,x_n)$.

An analogous result holds for Sobolev-Orlicz spaces of infinite order.

CHAN DYK VAN [1] - [3] also studied spaces of infinite order with weight in the strip $G = [0,a] \times \mathbb{R}^n$. In these spaces the metric is defined by

$$\varrho(u) = \sum_{n+|\alpha|=0}^{\infty} \int_G t^{q_n} |D_t^n D_x^\alpha u(t,x)|^{p_n}\, dx dt.$$

He proved that in the case $q_n/p_n \leqq$ const the criterion for nontriviality of the spaces with weight is the same as in the case of the spaces without weight.

In his work [1] A. JA. KOBILOV proved that the criterion for nontriviality of spaces $\mathring{W}^{\infty}\{a_\alpha, p_\alpha\}$ for a bounded domain, obtained in § 1, is also true for the case G = V, where V is a cone the angle of which is less than $\pi$. It is interesting to remark that for the halfspace $\mathbb{R}^+_{n+1} = \{t \geqq 0, x \in \mathbb{R}^n\}$ (the cone with angle $\pi$) the criterion for nontriviality has another character (see § 4, Chapter I).

The necessary and sufficient conditions for nontriviality of the spaces

$$W^{\infty}\{a_n, L\} = \left\{u(x): \sum_{n=0}^{\infty} a_n \| L^n u \|_{r_n}^{p_n} < \infty\right\},$$

where L is an elliptic operator of order 2m, were obtained by L. I. KLENINA [1]. She also studied more general spaces, the metric $\varrho(u)$ of which is defined as

$$\varrho(u) = \sum_{n=0} a_n \| A^n u \|^{p_n} < \infty ,$$

where A is a selfadjoint operator (in general unbounded) in Hilbert space. It turns out that the question of nontriviality of such spaces is closely connected with the spectrum of operator A. In particular, in the case of continuous spectrum and in the case of discrete spectrum there are results of the type of §§ 2, 3 of the present Chapter.

CHAPTER II

# ELLIPTIC BOUNDARY VALUE PROBLEMS OF INFINITE ORDER

## Introduction

In the first paragraphs of the present chapter we study the nonlinear Dirichlet problem of infinite order

$$L(u) \equiv \sum_{|\alpha|=0}^{\infty} (-1)^{|\alpha|} D^{\alpha} A_{\alpha}(x, D^{\gamma} u) = h(x), \quad x \in G,$$

$$D^{\omega} u \big|_{\Gamma} = 0, \quad |\omega| = 0, 1, \ldots,$$

the "energy" space of which is the space $\mathring{W}^{\infty}\{a_{\alpha}, p_{\alpha}\}$. As it was mentioned, this problem will be profound if the corresponding space $\mathring{W}^{\infty}\{a_{\alpha}, p_{\alpha}\}$ is nontrivial. In this connection we assume that the space $\mathring{W}^{\infty}\{a_{\alpha}, p_{\alpha}\}$ is nontrivial. By this assumption the solvability of the Dirichlet problem of infinite order is established. If the operator $L(u)$ is monotone, then the solution is unique.

The method of the investigation is the study of the limit behaviour of solutions of boundary value problems of order 2m (as $m \to \infty$)

$$L_{2m}(u_m) = h_m(x), \quad D^{\omega} u_m \big|_{\Gamma} = 0, \quad |\omega| < m,$$

where $L_{2m}(u_m)$ is a precise or slightly perturbed partial sum of the series $L(u)$. The limit element of the sequence $u_m(x)$ is a solution of the limit problem of infinite order $L(u) = h(x)$, $u(x) \in C_0^{\infty}(G)$.

Let us remark that, in contrast with the finite order Dirichlet problem, for the solvability of the Dirichlet problem of infinite order the monotonicity of the operator $L(u)$ is not required. In this respect nonlinear elliptic infinite order equations turn out to be related to systems of nonlinear algebraic equations, for the solvability of which, as is well known, the satisfaction of a coercivity condition is sufficient and no conditions of monotonicity are required (see, for example, J. L. LIONS [3], p. 66, JU. A. DUBINSKIJ [2], p. 7 et al.). Therefore, in the theory of nonlinear elliptic equations of infinite order the monotonicity condition is one of the possible conditions for the uniqueness of the solution, as takes place in the theory of linear elliptic problems of finite order or in the theory of algebraic equations.

Further, in § 3 the general problem of the behaviour of solutions of nonlinear boundary value problems

$$\sum_{|\alpha|=0}^{m} (-1)^{|\alpha|} D^{\alpha} A_{\alpha m}(x, D^{\gamma} u_m) = h_m(x), \quad D^{\omega} u_m|_{\Gamma} = 0, \qquad |\omega| < m,$$

is studied as $m \to \infty$.

We obtain the convergence of the family of solutions $u_m(x)$ to a solution of the limit equation, which may be either of infinite or finite order. One must remark that in the first case the passage to the limit as $m \to \infty$ is principally the same as in §§ 1, 2; but in the second case (the case of finite limit order) the foundations for the possibility of passing to the limit, as $m \to \infty$, is essentially more difficult, because in this case there is only the weak convergence $D^{\alpha} u_m \to D^{\alpha} u$ for $|\alpha| = r$, where r is the order of the limit equation. Here the technique of the theory of monotone operators (see J. MINTY [1], F. E. BROWDER [1] et al.) is developed for the case of a family of operators acting in spaces which also depend on the parameter m. In this connection we shall turn the reader's attention to a simple but usefull lemma (Lemma 3.1, a variant of the classical Fatou lemma) which is, from our point of view, of independent interest.

Let us also note example 3 which is connected with the question of the solvability of the noncoercive equation

$$\Delta u + \lambda^2 u = h(x), \quad x \in \mathbb{R}^n.$$

We also want to turn the reader's attention to the following problem: it is interesting to know an asymptotic behaviour of the passage to the limit $u_m(x) \to u(x)$, as $m \to \infty$. In one case (periodic conditions) this problem was solved by JU. A. KONJAEV [1]. In other cases the problem is open.

Let us note the work of N. ARONSZAJN [1] in which, from the position of polyharmonic analysis the behaviour of the boundary value problems, that are the reiterations of a selfadjoint boundary value problem of order 2m, was studied.

The fourth paragraph is devoted to the classical $L_2$-solutions of linear equations of infinite order

$$L(u) \equiv \sum_{|\alpha|=0}^{\infty} (-1)^{|\alpha|} a_{\alpha} D^{2\alpha} u(x) = h(x).$$

We say that the linear equation of infinite order is solvable in the classical sense if there exists a solution u(x) so that the series L(u) converges in the space $L_2$.

It turns out that the $L_2$-Dirichlet problem of infinite order is not correct in the classical sense, since for its solvability it is necessary that the right side h(x) should be orthogonal to some infinite set of functions.

On the contrary, the periodic problem and the problem of $L_2$-solvability in the full space $\mathbb{R}^n$ are correct in the classical sense.

Probably, the distinction of these results is not random but is connected with the nature of the operators of infinite order which are determined by entire or analytic functions. Let us note that all operators which occur in physics (for example, the translation operators) act on the periodic functions or on functions that are defined on the full space $\mathbb{R}^n$.

To conclude this paragraph we give a new method of solving some problems of mathematical physics. This method is based on the non-formal algebra of differential operators of infinite order.

In the last § 5 we consider one problem of the statistical theory of elasticity. The results of this paragraph are due to M. V. PAUKŠTO [1], who first connected the mechanical articles of V. V. NOVOŠILOV [1] and V. M. LEVIN [1] with the theory of boundary value problems of infinite order developed in the present Chapter.

## § 1. Dirichlet problem of infinite order (the main example)

Let $G \subset \mathbb{R}^n$ be a bounded domain with boundary $\Gamma$. In this domain the following nonlinear elliptic boundary value problem of infinite order

$$L(u) \equiv \sum_{|\alpha|=0}^{\infty} (-1)^{|\alpha|} D^{\alpha}(a_{\alpha}|D^{\alpha}u|^{p_{\alpha}-2} D^{\alpha}u) = h(x) \qquad (1.1)$$

$$D^{\omega}u|_{\Gamma} = 0, \quad |\omega| = 0,1,\dots \qquad (1.2)$$

is considered.

Here $a_{\alpha} = 0$, $p_{\alpha} > 1$ are any sequences of real numbers; moreover, the sequence $p_{\alpha}$ is bounded.

Obviously, the following "energy" space

$$\overset{\circ}{W}{}^{\infty}\{a_\alpha, p_\alpha\} = \left\{ u(x) \in C_0^\infty(G) : \varrho(u) \equiv \sum_{|\alpha|=0}^{\infty} a_\alpha \|D^\alpha u\|_{p_\alpha}^{p_\alpha} < \infty \right\}$$

corresponds to problem (1.1), (1.2).

The main assumption, which will be made throughout the following is the nontriviality of the space $\overset{\circ}{W}{}^{\infty}\{a_\alpha, p_\alpha\}$. In this space we shall find the solution of problem (1.1), (1.2).

The space $W^{-\infty}\{a_\alpha, p'_\alpha\}$ ($p'_\alpha = p_\alpha/(p_\alpha - 1)$) of the right sides h(x) is defined as the space which is formally dual to the space $\overset{\circ}{W}{}^{\infty}\{a_\alpha, p_\alpha\}$. Namely,

$$W^{-\infty}\{a_\alpha, p'_\alpha\} = \left\{ h(x) : h(x) = \sum_{|\alpha|=0}^{\infty} (-1)^{|\alpha|} a_\alpha D^\alpha h_\alpha(x) \right\},$$

where $h_\alpha(x) \in L_{p'}(G)$; moreover,

$$\varrho'(h) \equiv \sum_{|\alpha|=0}^{\infty} a_\alpha \|h_\alpha(x)\|_{p'_\alpha}^{p'_\alpha} < \infty . \tag{1.3}$$

Thus, the right side of (1.1) may be a generalized function with a singularity of infinite order.

By definition, the duality of the spaces $W^{-\infty}\{a_\alpha, p'_\alpha\}$ and $\overset{\circ}{W}{}^{\infty}\{a_\alpha, p_\alpha\}$ is given by the relation

$$\langle h, v \rangle = \sum_{|\alpha|=0}^{\infty} a_\alpha \int_G h_\alpha(x) \overline{D^\alpha v(x)}\, dx,$$

which, as it is not difficult to verify, is correct. Two elements $h_1 \in W^{-\infty}\{a_\alpha, p'_\alpha\}$ and $h_2 \in W^{-\infty}\{a_\alpha, p'_\alpha\}$ are considered as equal elements if

$$\langle h_1, v \rangle = \langle h_2, v \rangle ,$$

where $v(x) \in \overset{\circ}{W}{}^{\infty}\{a_\alpha, p_\alpha\}$ is an arbitrary function. It is easy to see that for every $u(x) \in \overset{\circ}{W}{}^{\infty}\{a_\alpha, p_\alpha\}$ we have

$$L(u) \in W^{-\infty}\{a_\alpha, p'_\alpha\},$$

that is

$$L(u) : \overset{\circ}{W}{}^{\infty}\{a_\alpha, p_\alpha\} \longrightarrow W^{-\infty}\{a_\alpha, p'_\alpha\}.$$

Definition 1.1. A function $u(x) \in \overset{\circ}{W}{}^{\infty}\{a_\alpha, p_\alpha\}$ is a solution of problem (1.1), (1.2) if for any function $v(x) \in \overset{\circ}{W}{}^{\infty}\{a_\alpha, p_\alpha\}$ the equality

$$\langle L(u),v\rangle \equiv \sum_{|\alpha|=0}^{\infty} (a_\alpha |D^\alpha u|^{p_\alpha - 2} D^\alpha u,\ D^\alpha v) = \langle h,v\rangle \qquad (1.4)$$

is valid. (Here $(u,v) \equiv \int_G u(x)\overline{v(x)}dx$.)

Theorem 1.1. For any $h(x) \in W^{-\infty}\{a_\alpha, p'_\alpha\}$ there exists one and only one solution $u(x) \in \mathring{W}^\infty\{a_\alpha, p_\alpha\}$ of problem (1.1), (1.2).

Proof. Let us consider the following sequence of nonlinear elliptic boundary value problems of order 2m (m = 0,1,...), which correspond to the partial sums of series (1.1):

$$L_{2m}(u_m) \equiv \sum_{|\alpha|=0}^{m} (-1)^{|\alpha|} D^\alpha(a_\alpha |D^\alpha u_m|^{p_\alpha - 2} D^\alpha u_m) = h_m(x), \qquad (1.5)_m$$

$$D^\omega u_m|_\Gamma = 0, \quad |\omega| < m, \qquad (1.6)_m$$

where

$$h_m(x) \equiv \sum_{|\alpha|=0}^{m} (-1)^{|\alpha|} a_\alpha D^\alpha h_\alpha(x).$$

It is well known (see M. I. VIŠIK [1], F. E. BROWDER [1], J. L. LIONS [3] et al.) that problem $(1.5)_m$, $(1.6)_m$ has a unique solution; moreover, the following inequality

$$\sum_{|\alpha|=0}^{m} a_\alpha \| D^\alpha u_m \|_{p_\alpha}^{p_\alpha} \leqq \sum_{|\alpha|=0}^{\infty} a_\alpha \| h_\alpha \|_{p'_\alpha}^{p'_\alpha} \leqq \varrho'(h) \qquad (1.7)$$

is valid.

From this, using the compact imbedding theorems of classical Sobolev spaces (see, for example, S. L. SOBOLEV [1]; S. M. NIKOLSKIJ [1]; O. V. BESOV, V. P. IL'IN and S. M. NIKOLSKIJ [1]) and a diagonal process, we obtain that the sequence $u_m(x)$ (a priori, a subsequence) converges uniformly in G with all its derivatives to a function $u(x) \in C_0^\infty(G)$[1]. Moreover, by virtue of (1.7) $u(x) \in \mathring{W}^\infty\{a_\alpha, p_\alpha\}$ and $\varrho(u) \leqq \varrho'(h)$.

Let us show that u(x) is the desired solution of problem (1.1), (1.2). For this we pass to the limit, as $m \to \infty$, in the problem $(1.5)_m$, $(1.6)_m$ in the sense of generalized functions on $\mathring{W}^\infty\{a_\alpha, p_\alpha\}$.

On the one hand (it is evident), we have

---

[1] Let us remark that every function $u_m(x)$ has an increasing but always finite smoothness; therefore, the convergence $u_m(x) \to u(x)$ in $C_0^\infty(G)$ means that $D^\alpha u_m(x) \to D^\alpha u(x)$ uniformly in G, beginning with an index m which depends on $\alpha$ and is large enough.

$$\lim_{m\to\infty} \langle L_{2m}(u_m),v\rangle = \lim_{m\to\infty} \langle h_m,v\rangle = \langle h,v\rangle , \tag{1.8}$$

where $v(x) \in \overset{\circ}{W}{}^{\infty}\{a_\alpha, p_\alpha\}$ is an arbitrary function.

On the other hand, let $m_0$ be a fixed number and $m > m_0$. We have

$$\langle L(u),v\rangle - \langle L_{2m}(u_m),v\rangle$$

$$\equiv \sum_{|\alpha|=0}^{m_0} (a_\alpha |D^\alpha u|^{p_\alpha - 2} D^\alpha u - a_\alpha |D^\alpha u_m|^{p_\alpha - 2} D^\alpha u_m, D^\alpha v)$$

$$+ \sum_{|\alpha|=m_0+1}^{\infty} (a_\alpha |D^\alpha u|^{p_\alpha - 2} D^\alpha u, D^\alpha v) - \sum_{|\alpha|=m_0+1}^{m} (a_\alpha |D^\alpha u_m|^{p_\alpha - 2} D^\alpha u_m, D^\alpha v) . \tag{1.9}$$

It is clear that the average term (it is equal to the remainder of the convergent series) is arbitrarily small for sufficiently large $m_0$. Moreover, taking into account the Young inequality and the estimate (1.7), we find (it is essential here that $p_\alpha$ is a bounded sequence) that for any $\varepsilon > 0$ there exists the inequality

$$\left| \sum_{|\alpha|=m_0+1}^{m} (a_\alpha |D^\alpha u_m|^{p_\alpha - 2} D^\alpha u_m, D^\alpha v) \right| \leqq \varepsilon \sum_{|\alpha|=m_0+1}^{m} a_\alpha \| D^\alpha u_m \|_{p_\alpha}^{p_\alpha}$$

$$+ K(\varepsilon) \sum_{|\alpha|=m_0+1}^{m} a_\alpha \| D^\alpha v \|_{p_\alpha}^{p_\alpha} \leqq \varepsilon \varrho'(h) + K(\varepsilon) \sum_{|\alpha|=m_0+1}^{\infty} a_\alpha \| D^\alpha v \|_{p_\alpha}^{p_\alpha} .$$

Therefore, this term is also arbitrary small if $m_0$ is large enough.

Finally, for any fixed $m_0$

$$\lim_{m\to\infty} \sum_{|\alpha|=0}^{m_0} (a_\alpha |D^\alpha u|^{p_\alpha - 2} D^\alpha u - a_\alpha |D^\alpha u_m|^{p_\alpha - 2} D^\alpha u_m, D^\alpha v) = 0,$$

since $u_m(x) \to u(x)$ uniformly in G together with all derivatives.

As a result, we shall have

$$\lim_{m\to\infty} \langle L_{2m}(u_m),v\rangle = \langle L(u),v\rangle$$

for any $v(x) \in \overset{\circ}{W}{}^{\infty}\{a_\alpha, p_\alpha\}$.

Comparing this relation with equality (1.8), we obtain that

$$\langle L(u),v\rangle = \langle h,v\rangle ,$$

where $v(x)$ is any function from the space $\overset{\circ}{W}{}^{\infty}\{a_\alpha, p_\alpha\}$. It means that $u(x)$ is a solution of problem (1.1), (1.2).

Let us prove the uniqueness of our solution. In order to do so let us obtain an estimate of the "modulus of continuity" of operator $L^{-1}$.

Proposition 1.1. For any functions $u(x) \in \mathring{W}^{\infty}\{a_\alpha, p_\alpha\}$ and $v(x) \in \mathring{W}^{\infty}\{a_\alpha, p_\alpha\}$ there are the following inequalities:

$$1) \quad \sum_{|\alpha|=0}^{\infty} a_\alpha \left( \int_{G_{+,\alpha}} |D^\alpha(u-v)|^{p_\alpha} dx + \frac{p_\alpha - 1}{2^{p_\alpha - 2}} \int_{G_{-,\alpha}} |D^\alpha(u-v)|^{p_\alpha - 2} \cdot |D^\alpha(u-v)|^2 dx \right) \leqq \langle L(u) - L(v), u - v \rangle ,$$

where $1 < p_\alpha < 2$, $G_{+,\alpha} = \{ x \in G: |D^\alpha(u+v)| \leqq |D^\alpha(u-v)| \}$, $G_{-,\alpha} = G \setminus G_{+,\alpha}$;

$$2) \quad \sum_{|\alpha|=0}^{\infty} a_\alpha \left( \frac{1}{2^{p_\alpha - 2}} \int_{G_{+,\alpha}} |D^\alpha(u-v)|^{p_\alpha} dx + 2 \int_{G_{-,\alpha}} |D^\alpha(u+v)|^{p_\alpha - 2} \cdot |D^\alpha(u-v)|^2 dx \right) \leqq \langle L(u) - L(v), u - v \rangle ,$$

where $p_\alpha \geqq 2$; $G_{+,\alpha}$ and $G_{-,\alpha}$ are the same as in inequality 1).

Proof. For any $m > 0$ we have

$$\langle L_{2m}(u) - L_{2m}(v), u - v \rangle \equiv \sum_{|\alpha|=0}^{m} (a_\alpha |D^\alpha u|^{p_\alpha - 2} D^\alpha u - a_\alpha |D^\alpha v|^{p_\alpha - 2} D^\alpha v, D^\alpha(u - v))$$

$$= \frac{1}{2} \sum_{|\alpha|=0}^{m} a_\alpha (p_\alpha - 1) \left( \int_{-1}^{1} \left| \frac{D^\alpha u + D^\alpha v}{2} + \tau \frac{D^\alpha u - D^\alpha v}{2} \right|^{p_\alpha - 2} d\tau \cdot D^\alpha(u - v), D^\alpha(u - v) \right) \geqq \sum_{|\alpha|=0}^{m} \frac{a_\alpha(p_\alpha - 1)}{2^{p_\alpha - 1}}$$

$$\cdot \left( \int_{-1}^{1} |q_\alpha + \tau|^{p_\alpha - 2} d\tau \, |D^\alpha(u - v)|^{p_\alpha - 1}, |D^\alpha(u - v)| \right), \tag{1.10}$$

where $q_\alpha \equiv |D^\alpha(u + v)| / |D^\alpha(u - v)|$. (Here the evident formula

$$\varphi(u) - \varphi(v) = \int_{-1}^{1} \varphi'_\tau \left( \frac{u + v}{2} + \tau \frac{u - v}{2} \right) d\tau ,$$

where $\varphi(t) \equiv |t|^{p_\alpha - 2} t$, was used.)

Computations show that for $1 < p < 2$

$$\int_{-1}^{1}|q+\tau|^{p-2}d\tau \geqq \begin{cases} \frac{2^{p-1}}{p-1}, & \text{if } |q| \leqq 1; \\ 2|q|^{p-2}, & \text{if } |q| > 1, \end{cases}$$

while for $p \geqq 2$

$$\int_{-1}^{1}|q+\tau|^{p-2}d\tau \geqq \begin{cases} \frac{2}{p-1}, & \text{if } |q| \leqq 1; \\ \frac{2^p}{p-1}|q|^{p-2}, & \text{if } |q| > 1. \end{cases}$$

Taking into account these inequalities and letting $m \to +\infty$ in (1.10), we obtain the desired inequalities. Proposition 1.1 is proved and hence the uniqueness of the solution of problem (1.1), (1.2) is proved too. Theorem 1.1 is completely proved.

Remark. In proposition 1.1 we do not require the boundedness of sequence $p_\alpha > 1$, that is , the sequence $p_\alpha$ may be arbitrary here.

Let us emphasize the following result which is a corollary of the proposition, proved just now.

Corollary 1.1. If $p_\alpha \geqq 2$, then the following inequality

$$\varrho\left(\frac{u-v}{2}\right) \leqq K\varrho'(L(u) - L(v)),$$

where $K > 0$ is a constant, is valid.

The latter inequality shows that under the condition $2 \leqq p_\alpha \leqq \text{const}$ the operator $L(u)$ defines a "homeomorphism" between the spaces $\mathring{W}^\infty\{a_\alpha, p_\alpha\}$ and $W^{-\infty}\{a_\alpha, p'_\alpha\}$.

In particular, if $p_\alpha \equiv p \geqq 2$, then the mapping

$$L(u):\ \mathring{W}^\infty\{a_\alpha, p\} \longrightarrow W^{-\infty}\{a_\alpha, p'\}, \tag{1.11}$$

$p' = p/(p-1)$, is a homeomorphism in the usual sense, that is a mapping which is one-to-one and continuous. Homeomorphism (1.11) is also true for $1 < p < 2$, but it is not a corollary of our estimates 1) and 2). This fact may be obtained if the mapping (1.11) is considered as a duality mapping of the space $\mathring{W}^\infty\{a_\alpha, p\}$ onto the conjugate space $W^{-\infty}\{a_\alpha, p'\}$ (see more precisely in Chapter IV).

Example. In the interval (a, b) the problem

$$\sum_{n=0}^{\infty}(-1)^n D^n\left(\frac{1}{(n!)^q}|D^n u|^{p-2}D^n u\right) = h(x), \tag{1.12}$$

$$D^n u(a) = D^n u(b) = 0, \quad n = 0,1,\dots \tag{1.13}$$

is considered. The corresponding space $\overset{\circ}{W}{}^{\infty}\left\{\frac{1}{(n!)^q}, p\right\}$ is nontrivial if and only if $p < q$. Therefore, if $p < q$ precisely, then problem (1.12), (1.13) has a unique solution for any $h(x) \in W^{-\infty}\left\{\frac{1}{(n!)^q}, p'\right\}$.

Counterexample. Let us consider the problem

$$\exp(-\Delta)u(x) = h(x), \quad x \in G,$$

$$D^{\omega}u|_{\Gamma} = 0, \quad |\omega| = 0,1,\dots$$

As it is not difficult to verify, the corresponding space $\overset{\circ}{W}{}^{\infty}\{\cdot, 2\}$ is trivial and hence this problem is reduced to the triviality $0 = 0$.

## § 2. Dirichlet problem of infinite order (the general case)

Let $G \subset \mathbb{R}^n$ be a bounded domain with boundary $\Gamma$. In $G$ we consider the following boundary value problem:

$$L(u) \equiv \sum_{|\alpha|=0}^{\infty} (-1)^{|\alpha|} D^{\alpha} A_{\alpha}(x, D^{\gamma}u) = h(x), \quad |\gamma| \leqq |\alpha|, \tag{2.1}$$

$$D^{\omega}u|_{\Gamma} = 0, \quad |\omega| = 0,1,\dots, \tag{2.2}$$

where $A_{\alpha}(x,\xi_{\gamma})$ are continuous functions of the arguments $x \in G$ and all possible $\xi_{\gamma}$.

Let the following conditions be fulfilled:

a) For $x \in G$ and for all possible $\xi_{\gamma}$ and $\eta_{\alpha}$, where $|\alpha| \leqq m$, $|\gamma| \leqq |\alpha|$ ($m \geqq 0$), the inequalities

$$\left| \sum_{|\alpha|=0}^{m} A_{\alpha}(x,\xi_{\gamma})\bar{\eta}_{\alpha} \right| \leqq K \sum_{|\alpha|=0}^{m} a_{\alpha} |\xi_{\alpha}|^{p_{\alpha}-1} |\eta_{\alpha}| + b$$

are valid. Here $K > 0$, $b \geqq 0$, $a_{\alpha} \geqq 0$, $p_{\alpha} > 1$ are some constants; moreover the sequence $p_{\alpha}$ is bounded.

b) The coercivity inequalities are satisfied, i. e.

$$\operatorname{Re} \sum_{|\alpha|=0}^{m} A_{\alpha}(x,\xi_{\gamma})\bar{\xi}_{\alpha} \geqq \delta \sum_{|\alpha|=0}^{m} a_{\alpha} |\xi_{\alpha}|^{p_{\alpha}} - C,$$

where $\delta > 0$, $C \geqq 0$ are constants.

c) The space $\overset{\circ}{W}{}^{\infty}\{a_\alpha, p_\alpha\}$ is nontrivial.

Theorem 2.1. Let the conditions a) - c) be satisfied. Then for any function $h(x) \in W^{-\infty}\{a_\alpha, p'_\alpha\}$ there exists at least one solution $u(x) \in \overset{\circ}{W}{}^{\infty}\{a_\alpha, p_\alpha\}$ of the problem (2.1), (2.2).

Remark. Before proving our theorem let us emphasize the prinzipal distinction between this theorem and the results of § 1. Namely, the equations

$$L(u) \equiv \sum_{|\alpha|=0}^{\infty} (-1)^{|\alpha|} D^\alpha(a_\alpha |D^\alpha u|^{p_\alpha - 2} D^\alpha u) = h(x)$$

are nonlinear equations of the monotone type, which (for the case of finite order) are now very well investigated. These equations are the basis of the modern monotonicity theory. On the contrary, in Theorem 2.1 no monotonicity conditions are required. Thus, for the solvability of strongly nonlinear problems of infinite order it is sufficient to have only coercivity conditions. In this relation the nonlinear problems of infinite order are similar to algebraic systems (m = 0), for which there is an analogous situation.

Proof. The idea of the proof ist the following.

First, the "truncated" equation of order 2m (the partial sum of the series (2.1)) is perturbed by a sufficiently "small" linear equation of order 2m + 2. The corresponding boundary value problem is always solvable. Then the passage to the limit is performed, as $m \to \infty$.

Thus, let us consider the sequence of the following Dirichlet problems of order 2m + 2 (m = 0,1,...):

$$\sum_{|\alpha|=m+1} (-1)^{m+1} c_\alpha D^{2\alpha} u_m + \sum_{|\alpha|=0}^{m} (-1)^{|\alpha|} D^\alpha A_\alpha(x, D^\gamma u_m) = h_m(x), \quad (2.3)_m$$

$$D^\omega u_m|_\Gamma = 0, \quad |\omega| \leq m, \qquad (2.4)_m$$

where $c_\alpha > 0$ are suitable small constants (see below),

$$h_m(x) = \sum_{|\alpha|=0}^{m} (-1)^{|\alpha|} a_\alpha D^\alpha h_\alpha(x).$$

The problem $(2.3)_m$, $(2.4)_m$ is a weakly nonlinear problem (the principal part of $(2.3)_m$ is linear). It is well known that such a problem is solvable (see, for example, J. L. LIONS [3], JU. A. DUBIN-

SKIJ [2] et al.); moreover, for solutions there are the estimates

$$\sum_{|\alpha|=m+1} c_\alpha \|D^\alpha u_m\|_2^2 + \sum_{|\alpha|=0}^{m} a_\alpha \|D^\alpha u_m\|_{p_\alpha}^{p_\alpha} \leqq K, \tag{2.5}$$

where $K = K(h) > 0$ is a constant.

From this (similar to § 1) we obtain that the sequence $u_m(x)$ (strictly speaking, a subsequence) converges uniformly in G, together with all derivatives, to a function $u(x) \in C_0^\infty(G)$; moreover, $u(x) \in \mathring{W}^\infty\{a_\alpha, p_\alpha\}$. This function is the desired solution of the problem (2.1), (2.2). To prove it let us choose the coefficients $c_\alpha > 0$, so that:

1) the space $\mathring{W}^\infty\{c_\alpha, 2\}$ is nontrivial,

2) there is the inclusion $\mathring{W}^\infty\{a_\alpha, p_\alpha\} \subset \mathring{W}^\infty\{c_\alpha, 2\}$

(the possibility of such a choice will be substantiated in Lemma 2.1 given below).

Then, using $(2.3)_m$, $(2.4)_m$, we shall have

$$\sum_{|\alpha|=m+1} c_\alpha (D^\alpha u_m, D^\alpha v) + \sum_{|\alpha|=0}^{m} (A_\alpha(x, D^\gamma u_m), D^\gamma v) = \langle h_m, v \rangle,$$

where $v(x) \in \mathring{W}^\infty\{a_\alpha, p_\alpha\}$ is an arbitrary function.

From this, using the condition a) and inequality (2.5) we obtain (in the same way, as in § 1) the identity

$$\langle L(u), v \rangle = \langle h, v \rangle, \quad v(x) \in \mathring{W}^\infty\{a_\alpha, p_\alpha\}, \tag{2.6}$$

which means that $u(x)$ is the solution of the problem (2.1), (2.2).

It only remains to choose the necessary sequence of $c_\alpha > 0$.

Lemma 2.1. For any nontrivial space $\mathring{W}^\infty\{a_\alpha, p_\alpha\}$ there exists a nontrivial space $\mathring{W}^\infty\{c_\alpha, 2\}$, so that

$$\mathring{W}^\infty\{a_\alpha, p_\alpha\} \subset \mathring{W}^\infty\{c_\alpha, 2\}. \tag{2.7}$$

Proof. (We shall produce the necessary numbers $c_\alpha > 0$.) Indeed, since the criterion for nontriviality of the spaces $\mathring{W}^\infty\{c_\alpha, 2\}$ is known, we may choose these numbers $c_\alpha > 0$, such that the space $\mathring{W}^\infty\{c_\alpha, 2\}$ is nontrivial. Consequently, it is necessary to choose these numbers such that there exists the inclusion (2.7).

For this we remark that for $p_\alpha \geqq 2$ there exists the inequality

$$\|D^\alpha u\|_2^2 \leqq K\|D^\alpha u\|_{p_\alpha}^2 \leqq K(\|D^\alpha u\|_{p_\alpha}^{p_\alpha} + 1),$$

where $K > 0$ is a constant, which does not depend on $\alpha$.

Hence, putting $c_\alpha \equiv a_\alpha$ (for all $\alpha$, such that $p_\alpha \geqq 2$), we shall have the inequality

$$\sum_{p_\alpha \geqq 2} c_\alpha \|D^\alpha u\|_2^2 \leqq K \sum_{p_\alpha \geqq 2} a_\alpha (\|D^\alpha u\|_{p_\alpha}^{p_\alpha} + 1) < \infty, \tag{2.8}$$

since $u(x) \in \mathring{W}^\infty\{a_\alpha, p_\alpha\}$.

Let us consider now the numbers $\alpha$, such that $p_\alpha < 2$. Since any function $u(x) \in \mathring{W}^\infty\{a_\alpha, p_\alpha\}$ is finite, then in conformity with the imbedding theorems of the classical Sobolev spaces

$$\|D^\alpha u\|_2^2 \leqq K \sum_{|\beta|=m} \|D^{\alpha+\beta} u\|_1 \leqq K \sum_{|\gamma|=N+m} \|D^\gamma u\|_1,$$

where $N = |\alpha|$, $m > \dim G$ is any number, $K = K$ (m, meas G) is a constant, not depending on $\alpha$.

Hence we immediately obtain the inequality

$$\|D^\alpha u\|_2 \leqq K \sum_{|\gamma|=N+m} \|D^\gamma u\|_{p_\gamma} \leqq K \sum_{|\gamma|=N+m} (\|D^\gamma u\|_{p_\gamma}^{p_\gamma} + 1).$$

Putting now $c_\alpha \equiv c_N$ (for all $\alpha$, such that $p_\alpha < 2$), we have the inequality

$$\sum_{p_\alpha < 2} c_\alpha^{1/2} \|D^\alpha u\|_2 \leqq K \sum_N \sum_{|\gamma|=N+m} c_N (\|D^\gamma u\|_{p_\gamma}^{p_\gamma} + 1). \tag{2.9}$$

It is clear that choosing $c_N \leqq a_\gamma$, $|\gamma| = N$, we can obtain the convergence of the series (2.9), since $u(x) \in \mathring{W}^\infty\{a_\alpha, p_\alpha\}$. The convergence of the series (2.9) implies the convergence of the series

$$\sum_{p_\alpha < 2} c_\alpha \|D^\alpha u\|_2^2.$$

The latter and the inequality (2.8) gives the proof of the lemma. Lemma 2.1 (and our theorem) is proved.

In conclusion, we point out one condition of the uniqueness of the solution of problem (2.1), (2.2).

Definition 2.1. The operator $L(u)$ is said to be a strongly monotone operator if for any functions $u(x) \in \mathring{W}^\infty\{a_\alpha, p_\alpha\}$, $v(x) \in \mathring{W}^\infty\{a_\alpha, p_\alpha\}$ the inequality

$$\operatorname{Re} \langle L(u) - L(v), u - v \rangle > 0, \; u \neq v,$$

is valid.

It is evident that in the case of a strongly monotone operator the solution of the problem (2.1), (2.2) is unique.

Example. Let us consider the problem

$$\sum_{|\alpha|=0}^{\infty} (-1)^{|\alpha|} D^{\alpha}(a_{\alpha}\varphi_{\alpha}(|D^{\alpha}u|)D^{\alpha}u) = h(x), \tag{2.7}$$

$$D^{\omega}u\big|_{\Gamma} = 0, \; |\omega| = 0,1,\ldots, \tag{2.8}$$

where $\varphi_{\alpha}(\xi)$ ($\xi \geqq 0$) are arbitrary continuous functions, such that

$$a|\xi|^{p_{\alpha}-2} \leqq \varphi_{\alpha}(\xi) \leqq b|\xi|^{p_{\alpha}-2}, \; 0 < a \leqq b, \; p_{\alpha} > 1.$$

If the space $\mathring{W}^{\infty}\{a_{\alpha}, p_{\alpha}\}$ is nontrivial, the problem (2.7), (2.8) is solvable. The solution is unique if the functions $\varphi_{\alpha}(\xi)$ are strongly monotone.

## § 3. The uniform correctness of a family of elliptic problems

Let $G \subset \mathbb{R}^n$ be a bounded region, its boundary being denoted by $\Gamma$. In G we consider the following family of nonlinear Dirichlet problems ($m = 1,2,\ldots$)

$$L_{2m}(u_m) \equiv \sum_{|\alpha|=0}^{m} (-1)^{|\alpha|} D^{\alpha}(a_{\alpha m}|D^{\alpha}u_m|^{p_{\alpha m}-2} D^{\alpha}u_m) = h_m(x), \tag{3.1$_m$}$$

$$D^{\omega}u_m\big|_{\Gamma} = 0, \quad |\omega| < m. \tag{3.2$_m$}$$

The behaviour of the solutions of problem $(3.1)_m$, $(3.2)_m$ is of interest to us. It turns out that two principal distinct cases must be considered: the case of an infinite order limit equation and the case of a finite order limit equation.

### I. The case of an infinite order limit equation.

Let us suppose that the following conditions are satisfied:

a) $a_{\alpha m} \geqq 0$, $p_{\alpha m} > 1$; moreover, the sequences $p_{\alpha m}$ are uniformly bounded sequences. Further, let $a_{\alpha m} \to a_{\alpha}$, $p_{\alpha m} \to p_{\alpha}$, as $m \to \infty$; moreover, $a_{\alpha} > 0$ for an infinite set of indices $\alpha$;

b) The space $\mathring{W}^{\infty}\{b_\alpha, q_\alpha\}$, where $b_\alpha = \sup_m a_{\alpha m}$, $q_\alpha = \sup_m p_{\alpha m}$, is nontrivial (it is easy to see that in this case the space $\mathring{W}^{\infty}\{a_\alpha, p_\alpha\}$ is also nontrivial; moreover, $\mathring{W}^{\infty}\{a_\alpha, p_\alpha\} \subset \mathring{W}^{\infty}\{b_\alpha, q_\alpha\}$).

c) The right sides $h_m(x)$ have the form

$$h_m(x) = \sum_{|\alpha|=0}^{m} (-1)^{|\alpha|} a_{\alpha m} D^\alpha h_{\alpha m}(x),$$

where $h_{\alpha m}(x) \in L_{p'_{\alpha m}}(G)$ and for any $m = 0,1,\ldots$

$$\sum_{|\alpha|=0}^{m} a_{\alpha m} \|h_{\alpha m}(x)\|_{p'_{\alpha m}}^{p'_{\alpha m}} \leq K, \ K > 0.$$

Finally, suppose that the sequence of the functions $h_m(x)$ converges to a function $h(x) \in W^{-\infty}\{a_\alpha, p'_\alpha\}$ in the sense that

$$\langle h_m, v\rangle \to \langle h, v\rangle \quad (m \to \infty)$$

for any function $v(x) \in \mathring{W}^{\infty}\{b_\alpha, q_\alpha\}$.

Definition 3.1. The family of boundary value problems $(3.1)_m$, $(3.2)_m$ is said to be uniformly correct if the corresponding sequence of solutions $u_m(x)$ converges (in the $C_0^\infty(G)$ sense) to a function $u(x) \in \mathring{W}^{\infty}\{a_\alpha, p_\alpha\}$; moreover, $u(x)$ is a solution of the limit problem

$$L(u) \equiv \sum_{|\alpha|=0}^{\infty} (-1)^{|\alpha|} D^\alpha(a_\alpha |D^\alpha u|^{p_\alpha - 2} D^\alpha u) = h(x), \tag{3.3}$$

$$D^\omega u|_\Gamma = 0, \quad |\omega| = 0,1,\ldots \tag{3.4}$$

Theorem 3.1. Let the conditions a) - c) be satisfied. Then the family of boundary value problems $(3.1)_m$, $(3.2)_m$ is uniformly correct.

Proof. The proof is essentially parallel to the proof of Theorems 1.1 and 2.1 of this Chapter. Namely, let us remark first of all that in view of condition c) there is the following estimate

$$\sum_{|\alpha|=0}^{m} a_{\alpha m} \|D^\alpha u_m\|_{p_{\alpha m}}^{p_{\alpha m}} \leq K, \tag{3.5}$$

where $u_m(x)$ are the solutions of $(3.1)_m$, $(3.2)_m$ and $K > 0$ is a constant.

Therefore, there is a subsequence (we do not change the notation) $u_m(x)$ which converges in $C_0^\infty(G)$ to a function $u(x)$. It follows from (3.5) that $u(x) \in \mathring{W}^{\infty}\{a_\alpha, p_\alpha\}$.

Let us show that $u(x)$ is a solution of the limit problem (3.3), (3.4).

First let $v(x)$ be a function from $\overset{\circ}{W}^{\infty}\{b_\alpha, q_\alpha\}$. Then, taking into account the condition c) we shall have

$$\langle L_{2m}(u_m), v\rangle = \langle h_m, v\rangle \to \langle h, v\rangle,$$

as $m \to \infty$.

On the other hand, if $m_0$ is a fixed number and $m > m_0$, then

$$\langle L(u), v\rangle - \langle L_{2m}(u_m), v\rangle \equiv \sum_{|\alpha|=m_0+1}^{\infty} (a_\alpha |D^\alpha u|^{p_\alpha - 2} D^\alpha u, D^\alpha v) -$$

$$- \sum_{|\alpha|=m_0+1}^{m} (a_{\alpha m} |D^\alpha u_m|^{p_{\alpha m}-2} D^\alpha u_m, D^\alpha v)$$

$$+ \sum_{|\alpha|=0}^{m_0} (a_\alpha |D^\alpha u|^{p_\alpha - 2} D^\alpha u - a_{\alpha m} |D^\alpha u_m|^{p_{\alpha m}-2} D^\alpha u_m, D^\alpha v) \equiv I_1 + I_2 + I_3.$$

Obviously, if $m_0$ is a sufficiently large number, then $I_1$ is arbitrarily small, since the corresponding series converges.

Further, in view of (3.5) we have the inequality

$$|I_2| \leq \varepsilon \sum_{|\alpha|=m_0+1}^{m} a_{\alpha m} \| D^\alpha u_m \|_{p_{\alpha m}}^{p_{\alpha m}} + K(\varepsilon) \sum_{|\alpha|=m_0+1}^{m} a_{\alpha m} \| D^\alpha v \|_{p_{\alpha m}}^{p_{\alpha m}}$$

$$\leq \varepsilon K + K_1(\varepsilon) \sum_{|\alpha|=m_0+1}^{\infty} b_\alpha \| D^\alpha v \|_{q_\alpha}^{p_{\alpha m}},$$

where $\varepsilon > 0$ is arbitrarily small and $K > 0$, $K_1(\varepsilon) > 0$ are some constants.

Due to this inequality and inclusion $v(x) \in \overset{\circ}{W}^{\infty}\{b_\alpha, q_\alpha\}$ we obtain that the value of $I_2$ is also arbitrarily small if $m_0$ is large enough.

Finally, for any fixed $m_0$ we have $I_3 \to 0$, as $m \to \infty$, since $u_m(x) \to u(x)$ in $C_0^\infty(G)$. As a result, we obtain that

$$\lim_{m\to\infty} \langle L_{2m}(u_m), v\rangle = \langle L(u), v\rangle \tag{3.7}$$

for any function $v(x) \in \overset{\circ}{W}^{\infty}\{b_\alpha, q_\alpha\}$.

Comparing (3.7) and (3.6) we immediately obtain that

$$\langle L(u), v\rangle = \langle h, v\rangle, \tag{3.8}$$

where $v(x) \in \mathring{W}^{\infty}\{b_\alpha, q_\alpha\}$ is an arbitrary function. Since the imbedding $\mathring{W}^{\infty}\{b_\alpha, q_\alpha\} \subset \mathring{W}^{\infty}\{a_\alpha, p_\alpha\}$ is dense, then the identity (3.8) holds for an arbitrary function $v(x) \in \mathring{W}^{\infty}\{a_\alpha, p_\alpha\}$. It follows that $u(x)$ is a solution of the limit problem (3.3), (3.4).

It remains to remark that $u(x)$ is the unique solution of the problem (3.3), (3.4) (the limit equation is strongly monotone) and, therefore, the whole sequence $u_m(x)$ converges to the solution $u(x)$. Theorem 3.1 is proved.

Let us emphasize the case $\mathring{W}^{\infty}\{b_\alpha, q_\alpha\} = \mathring{W}^{\infty}\{a_\alpha, p_\alpha\}$. (This equality holds if, for example, $a_{\alpha m} \to a_\alpha$, $p_{\alpha m} \to p_\alpha$ are decreasing and, in particular, if $a_{\alpha m} \equiv a_\alpha$, $p_{\alpha m} \equiv p_\alpha$. In the latter case the problems $(3.1)_m$, $(3.2)_m$ are the problems corresponding to the partial sums of the series (3.3).) In this case the nontriviality of the space $\mathring{W}^{\infty}\{a_\alpha, p_\alpha\}$ is not only a sufficient but a necessary condition of the uniform correctness of the family of problems $(3.1)_m$, $(3.2)_m$. Indeed, on the contrary, for any sequence $h_m(x)$ the solutions converge to zero by all means, as follows from inequality (3.5), and, therefore, cannot be a solution of the limit problem if the limit of the right side is not equal to zero identically. Consequently, there is the following.

Principle of the uniform correctness

Let the conditions a) - c) be satisfied and $\mathring{W}^{\infty}\{b_\alpha, q_\alpha\} = \mathring{W}^{\infty}\{a_\alpha, p_\alpha\}$. Then the family of elliptic problems is uniformly correct if and only if the space $\mathring{W}^{\infty}\{a_\alpha, p_\alpha\}$ is not empty, i. e. nontrivial.

In particular, we obtain that the Dirichlet problem of infinite order

$$\sum_{|\alpha|=0}^{\infty} (-1)^{|\alpha|} D^\alpha (a_\alpha |D^\alpha u|^{p_\alpha - 2} D^\alpha u) = h(x), \tag{3.9}$$

$$D^\omega u|_\Gamma = 0, \quad |\omega| = 0, 1, \ldots,$$

is uniformly correct if and only if the space $\mathring{W}^{\infty}\{a_\alpha, p_\alpha\}$ is nontrivial.

Remark. The previous results are valid for not only the Dirichlet problem but for any coercive problem; for example, the periodic problem ($x \in T^n$, where $T^n$ is the torus), the problem in the whole Euclidean space $\mathbb{R}^n$ etc.

Example 1. Let us consider the ordinary differential equation of infinite order

$$[\cos D]u(x) \equiv \sum_{n=0}^{\infty} \frac{(-1)^n}{(2n)!} D^{2n}u(x) = h(x), \tag{3.10}$$

where h(x) has the form

$$h(x) = \sum_{n=0}^{\infty} \frac{1}{(2n)!} D^n h_n(x);$$

moreover, $h_n(x) \in L_2$ and

$$\sum_{n=0}^{\infty} \frac{1}{(2n)!} \|h_n(x)\|_2^2 < \infty .$$

The corresponding partial sums have the form

$$\sum_{n=0}^{m} \frac{(-1)^n}{(2n)!} D^{2n}u_m(x) = h_m(x), \qquad (3.10)_m$$

where

$$h_m(x) = \sum_{n=0}^{m} \frac{1}{(2n)!} D^n h_n(x),$$

and, therefore, for equation $(3.10)_m$ any problem mentioned above is correct, i. e. Dirichlet problem, periodic problem and problem in $\mathbb{R}^n$. But for any finite interval the space $\mathring{W}^{\infty}\left\{\frac{1}{(2n)!}, 2\right\}$ is trivial and, consequently, the family of Dirichlet problems for $(3.10)_m$ is not uniformly correct, i. e. the Dirichlet problem of infinite order for equation (3.10) is not correct.

On the other hand, the space $W^{\infty}\left\{\frac{1}{(2n)!}, 2\right\}$ on the circle or on the whole line $\mathbb{R}^1$ is not empty and, therefore, the family of corresponding problems for the equations $(3.10)_m$ is uniformly correct. It follows that on the circle or on the whole line $\mathbb{R}^1$ the equation (3.10) has one and only one solution for any h(x).

Example 2. Let us consider the equation

$$(-\sqrt{I + \Delta})u(x) \equiv \sum_{K=0}^{\infty} a_K(-\Delta)^K u(x) = h(x), \tag{+}$$

where $a_K = (-1)^{K+1}\binom{K}{1/2} > 0$, $K = 0, 1, \ldots$

The function

$$\varphi(\xi) \equiv \sum_{K=0}^{\infty} a_K \xi^{2K}, \quad \xi^2 = \xi_1^2 + \ldots + \xi_n^2,$$

is the characteristic function of the corresponding functional space $W^{\infty}\{\cdot, 2\}$. This function $\varphi(\xi)$ is an analytic function in the unit sphere $|\xi| < 1$ and, therefore, the space $W^{\infty}\{\cdot, 2\}$ of function

$u(x): \mathbb{R}^n \to C^1$ is nontrivial (see Theorem 2.1, Chapter I). Consequently, the problem of the solvability of equation (*) in the whole Euclidean space $\mathbb{R}^n$ is correct. On the contrary, the periodic problem or Dirichlet problem of infinite order is noncorrect, since the corresponding "energy" spaces are empty.

Example 3. Similar to example 2 the operator

$$(I + \lambda^2\Delta)^{-1} \equiv \sum_{K=0}^{\infty} (-1)^K (\lambda^2\Delta)^K \quad (\lambda > 0)$$

maps isomorphically the space

$$W^{\infty}\{\lambda,\nabla\} = \left\{ u(x) \in C^{\infty}(\mathbb{R}^n) : \sum_{K=0}^{\infty} \lambda^{2K} \|\nabla^K u\|_2^2 < \infty \right\}$$

to the conjugate space $W^{-\infty}\{\lambda,\nabla\}$.

This example is interesting because the operator $I + \lambda^2\Delta$, as the inverse operator of the operator of infinite order $(I + \lambda^2\Delta)^{-1}$, maps the space $W^{-\infty}\{\lambda,\nabla\}$ to the space $W^{\infty}\{\lambda,\nabla\}$ isomorphically too, i. e. for any right side $h(x) \in W^{\infty}\{\lambda,\nabla\}$ the noncoercive elliptic equation

$$u + \lambda^2\Delta u = h(x), \quad \lambda > 0, \qquad (**)$$

has the unique solution

$$u(x) = (I + \lambda^2\Delta)^{-1} h(x) \equiv \sum_{K=0}^{\infty} (-1)^K \lambda^{2K} \Delta^K h(x)$$

from the space $W^{-\infty}\{\lambda,\nabla\}$.

It is known that the Laplace operator, as an operator in $L_2(\mathbb{R}^n)$, has the negative halfline $\mathbb{R}^1_-$ as a part of its continuous spectrum (the equation is not solvable for every function $h(x) \in L_2(\mathbb{R}^n)$). In view of that we should agree that the pair of spaces $W^{-\infty}\{\lambda,\nabla\}$, $W^{\infty}\{\lambda,\nabla\}$ is a natural pair of spaces for the mapping $I + \lambda^2\Delta$.

Let us emphasize that the equation (**) is considered in the sense of ultradistributions over the basic space $W^{\infty}\{\lambda,\nabla\}$. Therefore, in general such solutions are not solutions in the usual sense, that is in the sense of distributions over the classical basic space $\mathcal{D}$.

From Lemma 2.1, Chapter I, it follows that for any function $h(x) \in W^{\infty}\{\lambda,\nabla\}$ the Fourier transform $\tilde{h}(\xi)$ is concentrated in the sphere $|\xi| < \lambda^{-1}$. Consequently, we immediately obtain that the imbedding $W^{\infty}\{\lambda,\nabla\} \subset L_2(\mathbb{R}^n)$ is not dense. Thus, the operator $I + \lambda^2\Delta$, as an inverse operator to the operator $(I + \lambda^2\Delta)^{-1}$, is an essential expansion of the operator $I + \lambda^2\Delta$ in the sense of the usual generalized functions.

If the right side is such that the solution of equation (••)

$$u(x) = \sum_{K=0}^{\infty} (-1)^K \lambda^{2K} \Delta^K h(x)$$

is a convergent series (for example, in the sense of $L_2(\mathbb{R}^n)$), then $u(x)$ is a standard infinitely differentiable function and the relation

$$u + \lambda^2\Delta u = h(x)$$

is valid in the classical sense.

Our consideration of the equation (••) for $h(x) \in W^{\infty}\{\lambda,\nabla\}$ may be also supported by the possibility of solving it by the Fourier method. Indeed, as it was mentioned, any function $h(x) \in W^{\infty}\{\lambda,\nabla\}$ has Fourier transform $\tilde{h}(\xi)$ with support concentrated in the sphere $|\xi| < \lambda^{-1}$. Therefore, one may consider the regularization of the integral

$$u(x) = \frac{1}{(2\pi)^n} \int_{|\xi|<\lambda^{-1}} \frac{\tilde{h}(\xi)}{1 - \lambda^2\xi^2} \cdot e^{ix\xi} d\xi,$$

which gives a formal solution of equation (••). The consideration of $I + \lambda^2\Delta$ as the inverse operator to the infinite order operator $(I + \lambda^2\Delta)^{-1}$ is in fact one of the methods of regularization of the latter integral.

## II. The case of a finite order limit equation

We consider the problems $(3.1)_m$, $(3.2)_m$ again. We assume that the previous conditions a) - c) are satisfied, but we make the following additions to them. Namely, let the numbers $a_\alpha \equiv 0$ for all $\alpha$ such that $|\alpha| \geqq r + 1$, where $r \geqq 1$ is a natural number. Further, let us assume that the functions $h_m(x)$ have the form

$$h_m(x) = \sum_{|\alpha|=0}^{r} (-1)^{|\alpha|} D^\alpha h_{\alpha m}(x),$$

where $h_{\alpha m}(x) \in L_{s'_\alpha}(G)$, $s'_\alpha = \sup_m p'_{\alpha m}$, $p'_{\alpha m} = p_{\alpha m}/(p_{\alpha m}-1)$; moreover,

$$\| h_\alpha(x) - h_{\alpha m}(x) \|_{s'_\alpha} \to 0, \quad m \to \infty,$$

where $h_\alpha(x) \in L_{s'_\alpha}(G)$ are some certain functions.

Theorem 3.2. Suppose that all conditions mentioned above are satisfied. Then the sequence of the solutions of problems $(3.1)_m$, $(3.2)_m$ converges to a function u(x) weakly in the space $\mathring{W}^r_{\vec{s}}(G)$, where $\vec{s} = \left\{s_\alpha = \inf_m p_{\alpha m},\ |\alpha| \leq r\right\}$. In this connection $u(x) \in \mathring{W}^r_{\vec{p}}(G)$, where $\vec{p} = \left\{p_\alpha:\ |\alpha| \leq r\right\}$ and u(x) is a solution of the limit problem

$$L(u) \equiv \sum_{|\alpha|=0}^{r} (-1)^{|\alpha|} D^\alpha(a_\alpha |D^\alpha u|^{p_\alpha - 2} D^\alpha u) = h(x), \tag{3.12}$$

$$D^\omega u|_\Gamma = 0, \quad |\omega| < r, \tag{3.13}$$

where

$$h(x) = \sum_{|\alpha|=0}^{r} (-1)^{|\alpha|} D^\alpha h_\alpha(x).$$

Proof. For the proof we shall use the following lemma - a variant of the well known Fatou lemma.

Lemma. Let $u_m(x) \in L_{q_m}(G)$, $q_m \geq 1$, and $u_m(x) \to u(x)$ weakly in $L_1(G)$, i. e. for any bounded measurable function v(x)

$$(u_m, v) \to (u, v) \quad (m \to \infty).$$

Moreover, let $q_m \to q$, as $m \to \infty$, and $\underline{\lim}\, \|u_m\|_{q_m} < \infty$. Then $u(x) \in L_q(G)$ and

$$\|u\|_q \leq \underline{\lim}\, \|u_m\|_{q_m}. \tag{3.14}$$

Proof of lemma. Let v(x) be an arbitrary bounded measurable function. Then according to the conditions of the lemma

$$(u, v) = \lim_{m\to\infty} (u_m, v),$$

and, therefore,

$$|(u,v)| \leq \underline{\lim}\, (\|u_m\|_{q_m} \|v\|_{q'_m}) = \underline{\lim}\, \|u_m\|_{q_m} \cdot \lim_{m\to\infty} \|v\|_{q'_m}$$

$$= \underline{\lim}\, \|u_m\|_{q_m} \cdot \|v\|_{q'},$$

where $q' = q/(q - 1)$.

Let $v(x) = |u^N(x)|^{q-2} \cdot u^N(x)$, where $u^N(x) = 0$ if $|u(x)| \geq N$ and $u^N(x) = u(x)$ if $|u(x)| < N$ (N is a natural number). Then we immediately obtain that

$$|u^N|_q^q \leq \underline{\lim}\, |u_m|_{q_m} \cdot |u^N|_q^{q-1},$$

i. e.

$$|u^N|_q \leq \underline{\lim}\, |u_m|_{q_m}.$$

From this, as $N \to \infty$, we obtain that $u(x) \in L_q(G)$; moreover, inequality (3.14) holds. Q.E.D.

Let us turn to the proof of our theorem.

As before, for the solutions $u_m(x)$ there is the estimate

$$\sum_{|\alpha|=0}^{m} a_{\alpha m} \| D^\alpha u_m \|_{p_{\alpha m}}^{p_{\alpha m}} \leq K, \tag{3.15}$$

where $K > 0$ is a constant which does not depend on $m = 1,2,\ldots$

In particular, from (3.15) for all $|\alpha| \leq r$ we obtain that

$$\sum_{|\alpha|=0}^{r} a_\alpha \| D^\alpha u_m \|_{p_{\alpha m}}^{p_{\alpha m}} \leq K, \quad a_\alpha > 0,$$

and, consequently, we can consider that $D^\alpha u_m(x) \longrightarrow D^\alpha u(x)$ weakly in $L_{s_\alpha}(G)$ for all $|\alpha| \leq r$, where $s_\alpha = \inf\limits_m p_{\alpha m} > 1$. Due to the lemma, proved above (and it is essential)

$$u(x) \in \overset{\circ}{W}{}^r_{\vec{p}}(G), \text{ where } \vec{p} = \left\{ p_\alpha = \lim_{m\to\infty} p_{\alpha m},\ |\alpha| \leq r \right\}.$$

Let us show that $u(x)$ is the desired solution of the limit problem (3.12), (3.13). For this we remark first of all that by (3.15)

$$|D^\alpha u_m|^{p_{\alpha m}-2} D^\alpha u_m \in L_{p'_{\alpha m}}(G), \quad |\alpha| \leq r,$$

for all $m = 1,2,\ldots$; moreover,

$$\left\| |D^\alpha u_m|^{p_{\alpha m}-2} D^\alpha u_m \right\|_{p'_{\alpha m}} \leq K,$$

where the constant $K > 0$ depends on neither $\alpha$ nor $m$. Consequently, for $|\alpha| \leq r$ one can assume (once again choosing a subsequence, if necessary) that, as $m \to \infty$,

$$|D^\alpha u_m|^{p_{\alpha m}-2} D^\alpha u_m \longrightarrow g_\alpha(x)$$

weakly in $L_{q'_\alpha}(G)$, where $q'_\alpha = \inf\limits_m p'_{\alpha m}$. However, it is important to note that, in view of the lemma, $g_\alpha(x) \in L_{p'_\alpha}(G)$.

Taking into account everything said above, we obtain that for any function $v(x) \in \mathring{W}^{\infty}\{b_\alpha, q_\alpha\}$

$$\lim_{m\to\infty} \langle L_{2m}(u_m), v\rangle = \lim_{m\to\infty} \sum_{|\alpha|=r+1} (a_{\alpha m} |D^\alpha u_m|^{p_{\alpha m}-2} D^\alpha u_m, D^\alpha v)$$
$$+ \sum_{|\alpha|=0}^{r} (a_\alpha g_\alpha(x), D^\alpha v). \tag{3.16}$$

After the repetition of the considerations which were in the proof of Theorem 3.1 (estimate of $I_2$), we get that for all $|\alpha| > r$

$$\lim_{m\to\infty} (a_{\alpha m} |D^\alpha u_m|^{p_{\alpha m}-2} D^\alpha u_m, D^\alpha v) = 0.$$

Therefore, from (3.16) we have that

$$\lim_{m\to\infty} \langle L_{2m}(u_m), v\rangle = \sum_{|\alpha|=0}^{r} (a_\alpha g_\alpha(x), D^\alpha v).$$

On the other hand, from $(3.1)_m$, $(3.2)_m$ it evidently follows that

$$\lim_{m\to\infty} \langle L_{2m}(u_m), v\rangle = \langle h, v\rangle .$$

As a result, we obtain the identity

$$\langle h, v\rangle = \sum_{|\alpha|=0}^{r} (a_\alpha g_\alpha(x), D^\alpha v), \tag{3.17}$$

where $v(x) \in \mathring{W}^{\infty}\{b_\alpha, q_\alpha\}$ is an arbitrary function. In view of the density of the inclusion $\mathring{W}^{\infty}\{b_\alpha, q_\alpha\} \subset \mathring{W}^{r}_{\vec{p}}(G)$ this identity is also valid for any $v(x) \in \mathring{W}^{r}_{\vec{p}}(G)$ (let us recall that $h(x) \in W^{-r}_{\vec{p}'}(G)$, see (3.11)).

The rest of the proof essentially consists of the verification of the following assertion.

<u>Assertion</u>. For any function $v(x) \in W^{r}_{\vec{p}}(G)$ the following identity

$$\sum_{|\alpha|=0}^{r} (a_\alpha g_\alpha(x), D^\alpha v) = \sum_{|\alpha|=0}^{r} (a_\alpha |D^\alpha u|^{p_\alpha - 2} D^\alpha u, D^\alpha v) \tag{3.18}$$

is valid.

<u>Proof</u> (of the "monotonicity" method). Obviously, for any function $v(x) \in \mathring{W}^{\infty}\{b_\alpha, q_\alpha\}$ we have

$$\mathrm{Re} \sum_{|\alpha|=0}^{m} (a_{\alpha m} |D^\alpha u_m|^{p_{\alpha m}-2} D^\alpha u_m - a_{\alpha m}|D^\alpha v|^{p_{\alpha m}-2} D^\alpha v, D^\alpha (u_m - v)) \geq 0.$$

Since $\langle L_{2m}(u_m), u_m - v\rangle = \langle h_m, u_m - v\rangle$, then we obtain that

$$\mathrm{Re}\left[\langle h_m, u_m - v\rangle - \sum_{|\alpha|=0}^{m} (a_{\alpha m} |D^\alpha v|^{p_{\alpha m}-2} D^\alpha v, D^\alpha(u_m - v))\right] \geqq 0. \quad (3.19)$$

Let us pass to the limit in (3.19), as $m \to \infty$. For this we point out first of all that, in view of the conditions about behaviour of functions $h_m(x)$, we have

$$\langle h_m, u_m - v\rangle \longrightarrow \langle h, u - v\rangle . \quad (3.20)$$

Further, for $m > r$

$$I \equiv \sum_{|\alpha|=0}^{m} (a_{\alpha m} |D^\alpha v|^{p_{\alpha m}-2} D^\alpha v, D^\alpha(u_m - v))$$

$$= \sum_{|\alpha|=0}^{r} (a_{\alpha m} |D^\alpha v|^{p_{\alpha m}-2} D^\alpha v, D^\alpha(u_m - v))$$

$$+ \sum_{|\alpha|=r+1}^{m} (a_{\alpha m} |D^\alpha v|^{p_{\alpha m}-2} D^\alpha v, D^\alpha(u_m - v)) \equiv I_1 + I_2.$$

It is clear that

$$I_1 \longrightarrow \sum_{|\alpha|=0}^{r} (a_\alpha |D^\alpha v|^{p_\alpha - 2} D^\alpha v, D^\alpha(u - v)),$$

as $m \to \infty$.

Let us show that $I_2 \to 0$, as $m \to \infty$. In fact, if $r_0 > r+1$ is a fixed number, then for $m > r_0$

$$I_2 \equiv \sum_{|\alpha|=r+1}^{r_0} (a_{\alpha m} |D^\alpha v|^{p_{\alpha m}-2} D^\alpha v, D^\alpha(u_m - v))$$

$$+ \sum_{|\alpha|=r_0+1}^{m} (a_{\alpha m} |D^\alpha v|^{p_{\alpha m}-2} D^\alpha v, D^\alpha(u_m - v)) \equiv I_{21} + I_{22}.$$

Using the Young inequality and estimate (3.15), we obtain that for any $\varepsilon > 0$

$$|I_{22}| \leqq \varepsilon K + K(\varepsilon) \sum_{|\alpha|=r_0+1}^{m} a_{\alpha m} \|D^\alpha v\|_{p_{\alpha m}}^{p_{\alpha m}}$$

$$\leqq \varepsilon K + K_1(\varepsilon) \sum_{|\alpha|=r_0+1} b_\alpha (\|D^\alpha v\|_{q_\alpha}^{q_\alpha} + 1), \quad (3.21)$$

where (see the conditions of our theorem) $b_\alpha = \sup_m a_{\alpha m}$, $q_\alpha = \sup_m p_{\alpha m}$. Since $v(x) \in \mathring{W}^\infty\{b_\alpha, q_\alpha\}$, then it follows from (3.21) that $I_{22}$ is arbitrarily small, if $r_0$ is large enough. It only remains to remark that for any fixed $r_0$ the value $I_{22} \to 0$ due to $a_{\alpha m} \to 0$ for $|\alpha| > r$.

Taking into account these facts and the relation (3.20), we obtain from (3.19) that

$$\operatorname{Re}\left[\langle h, u - v\rangle - \sum_{|\alpha|=0}^{r} (a_\alpha |D^\alpha v|^{p_\alpha - 2} D^\alpha v, D^\alpha(u - v)\right] \geqq 0. \qquad (3.22)$$

The latter inequality was obtained for any function $v(x) \in \mathring{W}^\infty\{b_\alpha, q_\alpha\}$ but, as it was mentioned, the inclusion

$$\mathring{W}^\infty\{b_\alpha, q_\alpha\} \subset \mathring{W}^r_{\vec{p}}(G)$$

is dense and, consequently, relation (3.22) is valid for any function $v(x) \in \mathring{W}^r_{\vec{p}}(G)$.

The standard reasoning completes the proof. Namely, comparing (3.22) with (3.17) we have

$$\operatorname{Re} \sum_{|\alpha|=0}^{r} (a_\alpha g_\alpha(x) - a_\alpha |D^\alpha v|^{p_\alpha - 2} D^\alpha v, D^\alpha(u - v)) \geqq 0.$$

Putting here $v(x) = u(x) - \xi w(x)$, where $\xi > 0$, $w(x) \in \mathring{W}^r_{\vec{p}}(G)$ are arbitrary, we find, as $\xi \to +0$, that

$$\operatorname{Re} \sum_{|\alpha|=0}^{r} (a_\alpha g_\alpha(x) - a_\alpha |D^\alpha u|^{p_\alpha - 2} D^\alpha u, D^\alpha w) \geqq 0.$$

It follows immediately from this that

$$\sum_{|\alpha|=0}^{r} (a_\alpha g_\alpha(x) - a_\alpha |D^\alpha u|^{p_\alpha - 2} D^\alpha u, D^\alpha w) = 0. \quad \text{Q.E.D.}$$

Our assertion is proved. Together with the assertion Theorem 3.2 is also completely proved.

<u>Example</u>. In a region $G \subset \mathbb{R}^n$ the family of boundary value problems is considered:

$$(-1)^m a_m \Delta^m u_m(x) - \Delta u_m = h(x),$$

$$D^\omega u_m|_\Gamma = 0, \quad |\omega| < m,$$

where $a_m > 0$ $(m = 0, 1, \ldots)$ are certain constants, $h(x) \in W_2^{-1}(G)$. In conformity with the notations of this paragraph we have

$$a_{\alpha m} = \begin{cases} \dfrac{m!\, a_m}{\alpha!}, & |\alpha| = m; \\ 0, & |\alpha| < m; \end{cases}$$

further, $p_{\alpha m} = 2$; $r = 1$.

Hence, we have $b_\alpha = \sup_m a_{\alpha m} = \frac{|\alpha|!}{\alpha!} \cdot a_{|\alpha|}$. Thus, if the space $\overset{\circ}{W}{}^\infty\{b_\alpha, 2\}$ is nontrivial, then $u_m(x) \rightharpoonup u(x)$ weakly in $\overset{\circ}{W}{}^1_2(G)$, where $u(x) \in \overset{\circ}{W}{}^1_2(G)$ is a generalized solution of the Dirichlet problem

$$- \Delta u = h(x), \quad u|_\Gamma = 0.$$

Remark. In the proof of the main results of this paragraph we used a "nonstandard" condition b), i. e. the nontriviality of the space $\overset{\circ}{W}{}^\infty\{b_\alpha, q_\alpha\}$. Let us give an example, belonging to G. S. BALASHOVA, which shows that this condition is essential. Namely, let $u_m(x)$, $m = 0,1,\ldots,$ be the sequence of solutions of boundary value problems

$$L_{2m}(u_m) \equiv \frac{(-1)^m}{(2m)!} D^{2m}u_m - D^2u_m = 1 + f_m(x),$$

$$D^K u_m(0) = D^K u_m(1) = 0, \; 0 \leq K < m,$$

where $f_m(x) = - D^2\left[x^m(1 - x)^m\right]$.

For a given sequence of problems $b_n = 1/(2n)!$, $q_n = 2$, $n = 0,1,\ldots,$ and, therefore, the space $\overset{\circ}{W}{}^\infty\{b_n, q_n\}$ is trivial (see Theorem 1.1, Chapter I), i. e. the condition b) is not valid.

On the one hand, it is evident that $u_m(x) = x^m(1 - x)^m$ and, therefore, $u_m(x) \to 0$ uniformly on $[0,1]$ with all derivatives. On the other hand, the limit problem is

$$- D^2u = 1, \; u(0) = u(1) = 0$$

and its solution is $u(x) = \frac{1}{2}x(1 - x) \not\equiv 0$.

Thus, in spite of the fact that all conditions, except condition b), are satisfied, the assertion of Theorem 3.2 is false.

## § 4. Linear problems ($L_2$-theory)

Let us consider the following linear equation of infinite order

$$\sum_{|\alpha|=0}^{\infty} (-1)^{|\alpha|} a_\alpha D^{2\alpha}u(x) = h(x), \tag{4.1}$$

where $a_\alpha \geq 0$ are some constants.

It is evident that the equation (4.1) is a special case of the equation (1.1) ($p_\alpha \equiv 2$) considered in § 1. Therefore, for this equation one can consider all problems which were considered in § 1. Namely,

1) Dirichlet problem of infinite order;
2) Solvability in the whole space $\mathbb{R}^n$;
3) Periodic problem.

In conformity with the principle of uniform correctness (see § 3) these problems are correct, if the corresponding Sobolev spaces of infinite order, the metric of which is defined by the series

$$\sum_{|\alpha|=0}^{\infty} a_\alpha \|D^\alpha u\|_2^2 < \infty,$$

are nontrivial.

The solutions of these problems, as far as they exist, belong to the space $\mathring{W}^\infty\{a_\alpha, p_\alpha\}$ (or $W^\infty\{a_\alpha, p_\alpha\}$) and, therefore, they are infinitely differentiable functions. In spite of that they are generalized solutions, since they satisfy the equation (4.1) in the sense of distributions on the basic space $\mathring{W}^\infty\{a_\alpha, p_\alpha\}$ (or $W^\infty\{a_\alpha, p_\alpha\}$). It is evident, however, that in the linear case every term in the series (4.1) belongs to $L_2$, i. e. $a_\alpha D^{2\alpha} u(x) \in L_2$. Consequently, it is natural to consider the question of the convergence of the series (4.1) in the metric of the space $L_2$. This question is an analog of the question of the smoothness of the generalized solution of the equation of finite order or, what is the same, an analog of the classical solvability of the equation of finite order.

Definition 4.1. Let $h(x) \in L_2$. A generalized solution $u(x)$ of the problems 1) - 3) is called the classical $L_2$-solution if the equation (4.1) is satisfied in the $L_2$-sense, i. e. the series

$$\sum_{|\alpha|=0}^{\infty} (-1)^{|\alpha|} a_\alpha D^{2\alpha} u(x)$$

converges in the metric of the space $L_2$.

As will be seen below, the answers to this question are essentially different for the case of the Dirichlet problem of infinite order, on the one hand, and for the cases of $\mathbb{R}^n$ and of the torus $T^n$, on the other hand. Roughly speaking, the Dirichlet problem of infinite order is unsolvable in the classical sense and, in contrast, the equation (4.1) in the whole space $\mathbb{R}^n$ or in the torus $T^n$ has a classical solution.

Taking into account everything mentioned above we consider the following questions in this paragraph:

I. Dirichlet problem of infinite order;
II. Solvability in the whole space $\mathbb{R}^n$;
III. Variable coefficients;
IV. Absolute convergence;
V. Applications.

Let us consider the first question.

I. Dirichlet problem of infinite order.

In a bounded region $G \subset \mathbb{R}^n$ the Dirichlet problem of infinite order

$$L(u) \equiv \sum_{|\alpha|=0}^{\infty} (-1)^{|\alpha|} a_\alpha D^{2\alpha} u(x) = h(x), \quad x \in G, \tag{4.1}$$

$$D^{\omega} u \big|_{\Gamma} = 0, \quad |\omega| = 0, 1, \dots \tag{4.2}$$

is considered. Here, we recall, $a_\alpha \geq 0$ are constant coefficients, $h(x) \in L_2(G)$.

Let $u(x) \in \mathring{W}^{\infty}\{a_\alpha, p_\alpha\}$ be a generalized solution of the problem (4.1), (4.2). Let us suppose that this solution is the classical $L_2$-solution in the sense of Definition 4.1. In this case the equation (4.1) can be considered in the whole space $\mathbb{R}^n$, where $u(x)$ and $h(x)$ are extended by zero into $\mathbb{R}^n \setminus G$. Then, using the Fourier transform and taking into account its continuity in $L_2(\mathbb{R}^n)$, we immediately obtain that

$$a(\xi)\tilde{u}(\xi) = \tilde{h}(\xi),$$

where

$$a(\xi) \equiv \sum_{|\alpha|=0}^{\infty} a_\alpha \xi^{2\alpha}$$

is the symbol of the operator $L(u)$.

Since $u(x)$ is a finite function, then (Paley-Wiener theorem) its Fourier transform $\tilde{u}(\xi)$ admits an analytic extension into the full complex space $\mathbb{C}^n$, i. e. the function $\tilde{u}(\xi)$ can be extended into $\mathbb{C}^n$ as an entire function $\tilde{u}(z)$, $z = (z_1, \dots, z_n)$, $z_j \in \mathbb{C}^1$, $1 \leq j \leq n$. Further, according to the criterion for the nontriviality of the space $\mathring{W}^{\infty}\{a_\alpha, p_\alpha\}$ (see Ch. I, § 1) the function $a(\xi)$ is a power series with rapidly decreasing coefficients. Therefore, the function $a(\xi)$ can also be considered as an entire function $a(z)$.

Thus, the left side of the equality (4.3) admits an analytic extension $a(z)\tilde{u}(z)$ into $\mathbb{C}^n$. Thereby, the right side $\tilde{h}(\xi)$ has this property too, i. e. the function $\tilde{h}(\xi)$ is extended into $\mathbb{C}^n$ as an entire function; moreover,

$$a(z)\tilde{u}(z) = \tilde{h}(z), \quad z \in \mathbb{C}^n.$$

According to the well-known Picard theorem any entire function, which does not equal a polynomial, takes all values (with perhaps the exception of one value) infinitely many times. Since the coefficients $a_\alpha$ decrease rapidly, this exceptional value does not equal zero (see, for example, A. I. MARKUSHEVICH [1]). Thus, for the existence of the classical solution of the Dirichlet problem of infinite order it is necessary that the Fourier transform $\tilde{h}(\xi)$ of the function $h(x)$ should admit an analytic extension $\tilde{h}(z)$, which equals, obviously, zero in all points $z \in \mathbb{C}^n$, where $a(z) = 0$. Thereby, the question of classical solvability of the Dirichlet problem of infinite order is not a question of smoothness of the right side $h(x)$ (as it is in the elliptic theory of finite order) but is a question of the orthogonality of the right side $h(x)$ to some infinite set of certain functions.

It follows that the problem (4.1), (4.2) is not correct in the sense of Hadamard-Petrovskij. On the other hand, it is evident, that the function

$$u(x) = F^{-1}\left[\tilde{h}(\xi)/a(\xi)\right] \tag{4.4}$$

($F^{-1}$ is inverse Fourier transform) is the classical solution of the problem (4.1), (4.2), if all necessary conditions for the orthogonality are fulfilled so that the function $\tilde{h}(z)/a(z)$ is an entire function too. In addition, in view of Parseval's equality,

$$\int_G |L(u)|^2 dx = \int_{\mathbb{R}^n} a^2(\xi)\,|\tilde{u}(\xi)|^2 d\xi = \int_G |h(x)|^2 dx \tag{4.5}$$

and the series (4.1) converges in $L_2(G)$. Consequently, the formula (4.4) defines a classical $L_2$-solution. This solution is unique and it depends on the right side $h(x)$ continuously in the metric, defined by (4.5). Thus, the Dirichlet problem of infinite order is correct in the sense of TICHONOV (see A. N. TICHONOV [1], M. M. LAVRENTJEV [1], R. LATTES and J. L. LIONS [1] et al.).

II. Solvability in the whole space $\mathbb{R}^n$.

In the whole space $\mathbb{R}^n$ we consider the equation

$$L(u) \equiv \sum_{|\alpha|=0}^{\infty} (-1)^{|\alpha|} a_\alpha D^{2\alpha} u(x) = h(x), \quad a_0 > 0, \tag{4.6}$$

where $h(x) \in L_2(\mathbb{R}^n)$. We seek a classical $L_2$-solution in the sense of Definition 4.1.

According to the previous theory there exists one and only one generalized solution $u(x) \in W^\infty\{a_\alpha, 2\}$ $(\mathbb{R}^n)$ of the equation (4.6), i. e. the solution in the sense of the identity

$$\sum_{|\alpha|=0}^{\infty} (a_\alpha D^\alpha u, D^\alpha v) = (h,v), \quad v(x) \in W^\infty\{a_\alpha, 2\},$$

where $(\cdot,\cdot)$ means the integral in $\mathbb{R}^n$.

Our aim is to learn: may the series (4.6) converge in the space $L_2(\mathbb{R}^n)$? It turns out that the answer to this question depends on whether the characteristic function

$$a(\xi) \equiv \sum_{|\alpha|=0}^{\infty} a_\alpha \xi^{2\alpha} \tag{4.7}$$

(which is the symbol of operator $L(u)$) is an entire function or only an analytic function in a neighbourhood of zero.

In this connection let us recall (see § 2, Ch. I) that the space $W^\infty\{a_\alpha, 2\}$ $(\mathbb{R}^n)$ is nontrivial if and only if the function $a(\xi)$ is an analytic function at the point $\xi = 0$. In addition, two cases may take place:

1) the domain $G_a$ of convergence of series (4.7) is the whole space $\mathbb{R}^n$, i. e. $a(\xi)$ is an entire function;

2) the domain of convergence of series (4.7) is a domain $G_a \neq \mathbb{R}^n$.

Let us consider first the case $G_a = \mathbb{R}^n$. There is

Theorem 4.1. For any right side $h(x) \in L_2(\mathbb{R}^n)$ there exists one and only one classical $L_2$-solution of the equation (4.6). Moreover, $a(\xi)\tilde{u}(\xi) \in L_2(\mathbb{R}^n)$ and the equality

$$\int_{\mathbb{R}^n} a^2(\xi)|\tilde{u}(\xi)|^2 d\xi = \int_{\mathbb{R}^n} |\tilde{h}(\xi)|^2 d\xi \tag{4.8}$$

holds.

Proof. We show that the function

$$u(x) = F^{-1}\left[\frac{\tilde{h}(\xi)}{a(\xi)}\right],$$

where $F^{-1}$ is the inverse Fourier transform, is the desired solution. In fact, in view of Parseval's equality

$$\| h(x) - \sum_{|\alpha|=0}^{m} (-1)^{|\alpha|} a_\alpha D^{2\alpha} u \|_2^2 = \int_{\mathbb{R}^n} \left(1 - \frac{a_m(\xi)}{a(\xi)}\right)^2 |\tilde{h}(\xi)|^2 d\xi,$$

where

$$a_m(\xi) = \sum_{|\alpha|=0}^{m} a_\alpha \xi^{2\alpha}.$$

Since $0 \leqq a_m(\xi) \leqq a(\xi)$, then for any $\varepsilon > 0$

$$\| h(x) - \sum_{|\alpha|=0}^{m} (-1)^{|\alpha|} a_\alpha D^{2\alpha} u \|_2^2$$

$$\leqq \int_{|\xi|<R} \left(1 - \frac{a_m(\xi)}{a(\xi)}\right)^2 |\tilde{h}(\xi)|^2 d\xi + \int_{|\xi| \geqq R} |\tilde{h}(\xi)|^2 d\xi < \varepsilon ,$$

if $R$ and $m = m(R)$ are sufficiently large. It follows that the series (4.6) converges in the space $L_2(\mathbb{R}^n)$ to the function $h(x)$, i. e.

$$h(x) = \lim_{m\to\infty} \sum_{|\alpha|=0}^{m} (-1)^{|\alpha|} a_\alpha D^{2\alpha} u(x)$$

in the sense of $L_2(\mathbb{R}^n)$.

Thus, $u(x)$ is a classical $L_2$-solution of equation (4.6). The uniqueness and equality (4.8) are evident. The theorem is proved.

Let us consider now the case $G_a \neq \mathbb{R}^n$, i. e. the case where the characteristic function $a(\xi)$ is not an entire function.

We assert that in this case the series (4.6) converges in $L_2(\mathbb{R}^n)$ for only some functions $h(x) \in L_2(\mathbb{R}^n)$. Indeed, if the series (4.6) converges in the norm of $L_2(\mathbb{R}^n)$, then, applying Fourier transform, we have

$$\tilde{h}(\xi) = \lim_{m\to\infty} \sum_{|\alpha|=0}^{m} a_\alpha \xi^{2\alpha} \tilde{u}(\xi). \tag{4.9}$$

Recall (see § 2, Ch. I) that for any function $u(x) \in W^\infty\{a_\alpha, 2\}(\mathbb{R}^n)$ its Fourier transform $\tilde{u}(\xi)$ is concentrated in the domain of convergence $G_a$ of the characteristic function $a(\xi)$, i. e. supp $\tilde{u}(\xi) \subset G_a$. Consequently, it follows from the relation (4.9) that supp $\tilde{h}(\xi) \subset G_a$ too. Thus, the problem of the classical $L_2$-solvability may be valid only for a function $h(x) \in L_2(\mathbb{R}^n)$ the Fourier transform of which is concentrated in $G_a$. Thereby, in the case $G_a \neq \mathbb{R}^n$ the $L_2$-problem (4.6) has an infinite cokernel.

It is evident that the condition supp $\tilde{h}(\xi) \subset G_a$ is not only a necessary condition, but it is also a sufficient condition for the classical $L_2$-solvability of equation (4.6). Moreover, the solution is determined by the formula

$$u(x) = \frac{1}{(2\pi)^n} \int_{G_a} \frac{\tilde{h}(\xi)}{a(\xi)} e^{ix\xi} d\xi.$$

Remark. In the same way one can obtain that the periodic problem for equation (4.6) has always one and only one classical $L_2$-solution.

## III. Variable coefficients.

In this section we establish a theorem about solvability in $\mathbb{R}^n$ of the equation of infinite order

$$\sum_{|\alpha|=0}^{\infty} (-1)^{|\alpha|} a_\alpha(x) D^{2\alpha} u(x) = h(x), \tag{4.10}$$

where a (x) are certain variable coefficients.

For this we are in need of the definition of "convolutor" in the space $L_2(\mathbb{R}^n)$ (see L. HÖRMANDER [1], L. R. VOLEVICH and B. P. PANEJACH [1]).

Definition 4.2. A function $v(\xi)$ (in general, a generalized function) is called a convolutor in $L_2(\mathbb{R}^n)$ if for any function $u(\xi) \in L_2(\mathbb{R}^n)$ the convolution $v(\xi) * u(\xi)$ exists and belongs to $L_2(\mathbb{R}^n)$. Moreover, the map

$$u(\xi) \longrightarrow v(\xi) * u(\xi)$$

is continuous in $L_2(\mathbb{R}^n)$.

The latter in this definition means that there exists a constant $K = K(v)$ so that

$$\|v(\xi) * u(\xi)\|_2 \leq K \|u(\xi)\|_2,$$

where $u(\xi)$ is an arbitrary function. We shall denote the least of such constants as $\|v\|_{2\to2}$ and we shall call it the norm of the convolutor $v(\xi)$.

Let us now formulate the main theorem of this section.

Suppose that the coefficients $a_\alpha(x)$ equation (4.1) have the form

$$a_\alpha(x) = a_\alpha(1 + \varepsilon_\alpha(x)),$$

where $a_\alpha \geqq 0$ are certain numbers and $\varepsilon_\alpha(x)$ are certain functions. We suppose the following about these quantities:

1) the space $W^\infty\{a_\alpha, 2\}$ $(\mathbb{R}^n)$ is nontrivial;

2) the function $\tilde{\varepsilon}(\xi) = |\sup \tilde{\varepsilon}_\alpha(\xi)|$ is a convolutor in the space $L_2(\mathbb{R}^n)$ (here $\tilde{\varepsilon}_\alpha(\xi)$ means the Fourier transform of $\varepsilon_\alpha(x)$).

Finally, let us introduce the space of solutions of equation (4.10) as

$$H_a^\infty = \left\{ u(x): \ \|u\|^2_{a,\infty} = \int_{\mathbb{R}^n} a^2(\xi)|\tilde{u}(\xi)|^2 d\xi < \infty \right\} ,$$

where

$$a(\xi) \equiv \sum_{|\alpha|=0}^{\infty} a_\alpha \xi^{2\alpha}.$$

Theorem 4.2. Let conditions 1), 2) be satisfied and let the norm of the convolutor $\tilde{\varepsilon}(\xi)$ be less than one, i. e.

$$\|\tilde{\varepsilon}\|_{2\to 2} < 1.$$

Then for any function $h(x) \in L_2(\mathbb{R}^n)$ the equation (4.10) has one and only one solution $u(x) \in H_a^\infty$.

Proof. The proof of this theorem will be given in conformity with the well-known functional scheme of perturbation theory for inverse operators. Namely, we show that the equation (4.10) is equivalent to equation

$$v(x) + Tv(x) = h(x), \tag{4.11}$$

where T is a bounded operator in $L_2(\mathbb{R}^n)$. Consequently, if $\|T\| < 1$, then the equation (4.11) (and, therefore, the equation (4.10)) is one to one solvable.

The following assertion is the basis of our proof.

Assertion 4.1. Let $\tilde{\varepsilon}(\xi) \equiv \sup_\alpha |\tilde{\varepsilon}_\alpha(\xi)|$ be a convolutor in $L_2(\mathbb{R}^n)$. Then operator

$$L_\varepsilon(u) \equiv \sum_{|\alpha|=0}^{\infty} (-1)^{|\alpha|} a_\alpha \varepsilon_\alpha(x) D^{2\alpha} u(x) \tag{4.12}$$

is a continuous operator from $H_a^\infty$ to $L_2(\mathbb{R}^n)$; moreover, the norm $\|L_\varepsilon\|$ of this operator satisfies an inequality

$$\|L_\varepsilon\| \leqq \|\tilde{\varepsilon}\|_{2-2}.$$

Proof. Using Parseval's equality, we have

$$\|L_\varepsilon(u)\|_2 = \left\| \sum_{|\alpha|=0}^{\infty} (-1)^{|\alpha|} a_\alpha \tilde{\varepsilon}_\alpha * \widetilde{D^{2\alpha} u} \right\|_2$$

$$= \left\| \sum_{|\alpha|=0}^{\infty} (-1)^{|\alpha|} a_\alpha \int_{\mathbb{R}^n} \tilde{\varepsilon}_\alpha(\eta)(\xi-\eta)^{2\alpha} \tilde{u}(\xi-\eta)\, d\eta \right\|_2$$

$$\leq \left\| \sum_{|\alpha|=0}^{\infty} a_\alpha \int_{\mathbb{R}^n} |\varepsilon_\alpha(\eta)|\, (\xi-\eta)^{2\alpha} |\tilde{u}(\xi-\eta)|\, d\eta \right\|_2$$

$$\leq \left\| \sum_{|\alpha|=0}^{\infty} a_\alpha \int_{\mathbb{R}^n} \tilde{\varepsilon}(\eta)(\xi-\eta)^{2\alpha} |\tilde{u}(\xi-\eta)|\, d\eta \right\|_2$$

$$= \left\| \int_{\mathbb{R}^n} \tilde{\varepsilon}(\eta) a(\xi-\eta) |\tilde{u}(\xi-\eta)|\, d\eta \right\|_2 =$$

$$= \left\| \tilde{\varepsilon}(\xi) * a(\xi) |\tilde{u}(\xi)| \right\|_2 \leq \|\tilde{\varepsilon}\|_{2\to 2} \|u\|_{a,\infty},$$

since, by hypothesis, $\tilde{\varepsilon}(\xi)$ is a convolutor in $L_2(\mathbb{R}^n)$ and

$$\|a(\xi)|\tilde{u}(\xi)|\|_2 = \|u\|_{a,\infty}.$$

These calculations completely prove our assertion.

Let us now turn to the proof of the theorem. For this let us represent the equation (4.10) in the form

$$L_0(u) + L_\varepsilon(u) = h(x), \tag{4.13}$$

where $L_\varepsilon(u)$ is determined by formula (4.12) and

$$L_0(u) \equiv \sum_{|\alpha|=0}^{\infty} (-1)^{|\alpha|} a_\alpha D^{2\alpha} u(x).$$

According to Theorem 4.1 operator

$$L_0(u): H_a^\infty \longrightarrow L_2(\mathbb{R}^n)$$

is an isometric isomorphism. Consequently, making the replacement

$$u(x) = L_0^{-1} v(x), \quad v(x) \in L_2(\mathbb{R}^n),$$

in the equation (4.13), we obtain an equivalent equation in $L_2(\mathbb{R}^n)$

$$v(x) + Tv(x) = h(x), \tag{4.11}$$

where $T = L_\varepsilon L_0^{-1}$.

According to assertion 4.1 the operator $L_\varepsilon(u)$ is a continuous operator in $L_2(\mathbb{R}^n)$ and, consequently,

$$\|T\| \leqq \|L_\varepsilon\|_{H_a^\infty \to L_2} \cdot \|L_0^{-1}\|_{L_2 \to H_a^\infty}$$

$$\leqq \|\tilde{\varepsilon}\|_{2\to 2} \cdot \|L_0^{-1}\|_{L_2 \to H_a^\infty} = \|\tilde{\varepsilon}\|_{2\to 2} < 1.$$

Thus, due to a well-known theorem of functional analysis, for any function $h(x) \in L_2(\mathbb{R}^n)$ the equation (4.11) has a unique solution $u(x) \in H_a^\infty$. The theorem is proved.

Remark. The application of the theorem just proved rests obviously on the verification of the fact that $\tilde{\varepsilon}(\xi) = \sup_\alpha |\tilde{\varepsilon}_\alpha(\xi)|$ is a convolutor in $L_2(\mathbb{R}^n)$. The assumption $\tilde{\varepsilon}(\xi) \in L_1(\mathbb{R}^n)$ is a simple sufficient condition for the validity of condition 2). Indeed, in this case, as it is known, for any function $u(\xi) \in L_2(\mathbb{R}^n)$ the convolution $\tilde{\varepsilon} * u \in L_2(\mathbb{R}^n)$ too; moreover,

$$\|\tilde{\varepsilon} * u\| \leqq \|\tilde{\varepsilon}\|_1 \cdot \|u\|_2.$$

## IV. Absolute convergence.

In this paragraph we consider the question of the absolute convergence in $L_2(\mathbb{R}^n)$ of the series

$$L(u) \equiv \sum_{|\alpha|=0}^{\infty} (-1)^{|\alpha|} a_\alpha D^{2\alpha} u(x),$$

where $a_\alpha \geqq 0$ are constant coefficients.

Let $h(x) \in L_2(\mathbb{R}^n)$ and let $u(x) \in H_a^\infty$ be a classical $L_2$-solution of equation $L(u) = h(x)$. Simple calculations (with the use of Parseval's equality) show that

$$\sum_{|\alpha|=0}^{\infty} a_\alpha \|D^{2\alpha} u\|_2 < \infty$$

if and only if the function $h(x) \in L_2(\mathbb{R}^n)$ satisfies, in addition, the following condition

$$\sum_{|\alpha|=0}^{\infty} a_\alpha \|\xi^{2\alpha} a^{-1}(\xi) \tilde{h}(\xi)\|_2 < \infty, \qquad (4.14)$$

where, we recall, $a(\xi)$ is the symbol of the operator $L(u)$.

Finally, we give an example of a function $h(x) \in L_2(\mathbb{R}^n)$ for which the series (4.14) diverges. Namely, we put

$$a_n = \frac{1}{n!}, \quad a(\xi) = \sum_{n=0}^{\infty} \frac{1}{n!} \xi^{2n} = e^{\xi^2}.$$

Further, let

$$\tilde{h}(\xi) = \sum_{n=2}^{\infty} \tilde{h}_n(\xi),$$

where

$$\tilde{h}_n(\xi) = \begin{cases} \dfrac{1}{n^{1/4}\ln n} & \text{for } x \in (\sqrt{n}, \sqrt{n+1}), \\ 0 & \text{for } x \bar{\in} (\sqrt{n}, \sqrt{n+1}). \end{cases}$$

Note that the function $\xi^{4n} e^{-2\xi^2}$ has its maximal value (only one for $\xi > 0$) at the point $\xi = \sqrt{n}$ so on the interval $(\sqrt{n}, \sqrt{n+1})$ the inequality

$$\xi^{4n} e^{-2\xi^2} \geqq (n+1)^{2n} e^{-2(n+1)}$$

is valid. From this we obtain that

$$\| \xi^{2n} e^{-\xi^2} \tilde{h}_n(\xi) \|_2^2 = \int_{\sqrt{n}}^{\sqrt{n+1}} \xi^{4n} e^{-2\xi^2} |\tilde{h}_n(\xi)|^2 d\xi$$

$$\geqq \frac{(n+1)^{2n}}{e^{2(n+1)}} \cdot \frac{\sqrt{n+1} - \sqrt{n}}{n^{1/2}\ln^2 n}.$$

Consequently,

$$\sum_{n=2}^{\infty} a_n \| \xi^{2n} a^{-1}(\xi)\tilde{h}(\xi) \|_2 \geqq \sum_{n=2}^{\infty} \frac{(n+1)^n e^{-(n+1)}}{n!\, n^{1/4}\ln n} (\sqrt{n+1} - \sqrt{n})^{\frac{1}{2}}$$

$$\sim \sum_{n=2}^{\infty} \frac{e^n}{\sqrt{n}\, n^n} \cdot \frac{(n+1)^n e^{-(n+1)}}{n^{1/4} \ln n} \cdot \frac{1}{n^{1/4}} \sim \sum_{n=2}^{\infty} \frac{1}{n \ln n} = +\infty,$$

i. e. the series (4.14) diverges. Thus, the function $h(x) = F^{-1}\tilde{h}(\xi)$ does not satisfy the condition (4.14).

On the other hand,

$$\| h(x) \|_2^2 = \int_{-\infty}^{\infty} |\tilde{h}(\xi)|^2 d\xi = \sum_{n=2}^{\infty} \int_{\sqrt{n}}^{\sqrt{n+1}} |\tilde{h}_n(\xi)|^2 d\xi =$$

$$= \sum_{n=2}^{\infty} \frac{1}{n^{1/2}\ln^2 n} (\sqrt{n+1} - \sqrt{n}) \sim \sum_{n=2}^{\infty} \frac{1}{n \ln^2 n} < \infty,$$

i. e. $h(x) \in L_2(\mathbb{R}^n)$.

Summary. The space of right sides h(x), for which the equation (4.10) is solvable in the sense of absolute convergence, is not the whole space $L_2(\mathbb{R}^n)$ but is only the subspace defined by the additional condition (4.14).

## V. Applications.

In this paragraph we show that the previous results can be applied to some problems of mathematical physics.

1. Boundary value problem for the Laplace equation in the strip. Let $G = \left\{ -\frac{1}{2} < y < \frac{1}{2},\ x \in \mathbb{R}^1 \right\}$ be the strip of the variables y, x. We consider the following boundary value problem

$$\frac{\partial^2 u}{\partial y^2} + \frac{\partial^2 u}{\partial x^2} = 0, \qquad (4.15)$$

$$u(\tfrac{1}{2}, x) = \varphi_+(x), \qquad (4.16)$$

$$\frac{\partial u}{\partial y}(-\tfrac{1}{2}, x) = \varphi_-(x), \qquad (4.17)$$

where $\varphi_\pm(x) : \mathbb{R}^1 \to \mathbb{C}^1$ are functions.

We solve (for the time being only formally) this problem by the following method. Namely, let us put $p = i\partial/\partial x$ and consider the equation (4.15) as an ordinary differential equation

$$\frac{\partial^2 u}{\partial y^2} - p^2 u = 0,\ y \in (-\tfrac{1}{2}, \tfrac{1}{2}). \qquad (4.18)$$

The general solution of (4.18) is

$$u(y,x) = e^{yp}\, c_1(x) + e^{-yp}\, c_2(x),$$

where $c_1(x)$, $c_2(x)$ are arbitrary functions.

These functions are determined by the conditions (4.16), (4.17), which give the following system of algebraic equations.

$$e^{\frac{p}{2}}\, c_1(x) + e^{-\frac{p}{2}}\, c_2(x) = \varphi_+(x)$$

$$p\, e^{-\frac{p}{2}}\, c_1(x) - p\, e^{\frac{p}{2}}\, c_2(x) = \varphi_-(x).$$

Solving this system, we obtain that the solution $u(y,x)$ is expressed by the formula

$$u(y,x) = \left[ \cos(y + \tfrac{1}{2})\, ip \right] u_+(x) + \left[ \frac{\sin(y - \frac{1}{2})\, ip}{ip} \right] u_-(x),$$

where the functions $u_\pm(x)$ satisfy the relation

$$\cos ip \cdot u_\pm(x) = \varphi_\pm(x),\ x \in \mathbb{R}^1.$$

Replacing in these relations the symbol p by $i\partial/\partial x$, we finally obtain the formula

$$u(y,x) = \left[\cos(y+\tfrac{1}{2})\frac{\partial}{\partial x}\right]u_+(x) + \left[\frac{\sin(y-\frac{1}{2})\frac{\partial}{\partial x}}{\frac{\partial}{\partial x}}\right]u_-(x), \qquad (4.19)$$

where the functions $u_\pm(x)$ are the solutions of the equations

$$\left[\cos\frac{\partial}{\partial x}\right]u_\pm(x) = \varphi_\pm(x), \quad x\in\mathbb{R}^1. \qquad (4.20)$$

Let $\varphi_\pm(x)\in L_2(\mathbb{R}^1)$. Then (in conformity with Theorem 4.1) the equations (4.20) have the unique solutions $u_\pm(x)$ such that $\operatorname{ch}\xi\cdot\tilde{u}_\pm(\xi)\in L_2(\mathbb{R}^1)$; moreover,

$$\int_{-\infty}^{\infty}\operatorname{ch}^2\xi|\tilde{u}_\pm(\xi)|^2d\xi = \int_{-\infty}^{\infty}|\tilde{\varphi}_\pm(\xi)|^2d\xi.$$

It is easy to see in this case that for any $y\in(-\frac{1}{2},\frac{1}{2})$ the formula (4.19) gives a twice differentiable function $u(y,x)$, i. e. this function is the classical solution of the initial problem (4.15)-(4.17).

If the boundary functions $\varphi_\pm(x)\in W^{-\infty}\left\{\frac{1}{2n!},2\right\}(\mathbb{R}^1)$, then the equations (4.20) have the generalized solutions

$$u_\pm(x)\in W^{\infty}\left\{\frac{1}{2n!},2\right\}(\mathbb{R}^1)$$

and, consequently, the function $u(y,x)$ is also the generalized solution in the basic space $W^{\infty}\left\{\frac{1}{2n!},2\right\}(\mathbb{R}^1)$ of the problem (4.15)-(4.17).

2. Cauchy problem for the wave equation. In the halfplane $\mathbb{R}^2_+ = \left\{t>0,\ x\in\mathbb{R}^1\right\}$ we consider the initial problem

$$\frac{\partial^2u}{\partial t^2} - \frac{\partial^2u}{\partial x^2} = 0, \qquad (4.21)$$

$$u(0,x) = \varphi(x), \qquad (4.22)$$

$$\frac{\partial u}{\partial t}(0,x) = \psi(x), \qquad (4.23)$$

where the functions $\varphi(x)$ and $\psi(x)$ are given.

Putting $p = i\partial/\partial x$, we have from (4.21)

$$\frac{\partial^2u}{\partial t^2} + p^2u = 0.$$

From this we obtain the formula

$$u(t,x) = e^{itp}\, c_1(x) + e^{-itp}\, c_2(x),$$

where the functions $c_1(x)$, $c_2(x)$ are arbitrary. In order to determine these functions we use the initial conditions (4.22), (4.23), from which

$$c_1(x) + c_2(x) = \varphi(x),$$

$$ip\, c_1(x) - ip\, c_2(x) = \psi(x).$$

After elementary calculations we get the formula

$$u(t,x) = \frac{e^{itp} + e^{-itp}}{2}\,\varphi(x) + \frac{e^{itp} - e^{-itp}}{2\,ip}\,\psi(x).$$

Taking into account the relation $p = i\partial/\partial x$, we find that the desired solution has the form

$$u(t,x) = \frac{e^{t\frac{\partial}{\partial x}} + e^{-t\frac{\partial}{\partial x}}}{2}\,\varphi(x) + \frac{e^{t\frac{\partial}{\partial x}} - e^{-t\frac{\partial}{\partial x}}}{2\frac{\partial}{\partial x}}\,\psi(x) \tag{4.24}$$

or, equivalently,

$$u(t,x) = \left[\operatorname{ch} t\frac{\partial}{\partial x}\right]\varphi(x) + \left[\frac{\operatorname{sh} t\frac{\partial}{\partial x}}{\frac{\partial}{\partial x}}\right]\psi(x). \tag{4.25}$$

Let us elucidate what the initial functions $\varphi(x)$ and $\psi(x)$ should be in order that the formulae (4.24) and (4.25) have a nonformal sense. For this we consider the space $W^{\infty}$ of functions $\varphi(x)$ such that for any $t > 0$

$$\varrho(t,\varphi) \equiv \sum_{n=0}^{\infty} \frac{t^n}{n!}\,\|D^n\varphi\|_2 < \infty, \quad D \equiv d/dx,$$

i. e. we consider the space

$$W^{\infty} = \left\{\varphi(x) \in C^{\infty}(\mathbb{R}^1) : \varrho(t,\varphi) < \infty \quad , \quad \forall t > 0\right\},$$

where $\|\cdot\|_2$ is the norm in $L_2(\mathbb{R}^1)$. (The nontriviality of this space follows from Theorem 2.1, Ch. I.)

It is obvious that the operator

$$e^{t\frac{\partial}{\partial x}} \equiv \sum_{n=0}^{\infty} \frac{t^n}{n!}\,\frac{\partial^n}{\partial x^n}$$

is determined in the sense of $L_2(\mathbb{R}^1)$ for any $\varphi(x) \in W^\infty$; moreover,

$$e^{t\frac{\partial}{\partial x}} \varphi(x) \equiv \sum_{n=0}^{\infty} \frac{t^n}{n!} D^n\varphi(x) = \varphi(x+t), \tag{4.26}$$

since any function $\varphi(x)$ from the space $W^\infty$ is, obviously, an entire function.

Using the standard techniques of $\delta$-type sequences, it is easy to prove that the imbedding $W^\infty \subset L_2(\mathbb{R}^1)$ is everywhere dense. Consequently, the right side of the equality (4.26) admits closure up to an arbitrary function $\varphi(x) \in L_2(\mathbb{R}^1)$. It follows that the left side of (4.26) admits closure up to an arbitrary function $\varphi(x) \in L_2(\mathbb{R}^1)$ too. Therefore, the operator $\exp(t\partial/\partial x)$ has a closed extension (we shall note this extension as $\exp(t\partial/\partial x)$ too), which acts as a translation operator:

$$e^{t\frac{\partial}{\partial x}} \varphi(x) = \varphi(x+t), \quad \varphi(x) \in L_2(\mathbb{R}^1).$$

It follows, consequently, that the first term in the formulae (4.24), (4.25) has a nonformal sense, namely

$$\frac{e^{t\frac{\partial}{\partial x}} + e^{-t\frac{\partial}{\partial x}}}{2} \varphi(x) = \frac{\varphi(x+t) + \varphi(x-t)}{2}.$$

Putting

$$\left(\frac{\partial}{\partial x}\right)^{-1} \psi(x) = \int_0^x \psi(\xi)\,d\xi + c,$$

where c is an arbitrary constant, we get that the second term in the formulae (4.24), (4.25) is also determined for any function $\psi(x) \in L_2(\mathbb{R}^1)$; moreover,

$$\frac{e^{t\frac{\partial}{\partial x}} - e^{-t\frac{\partial}{\partial x}}}{2\frac{\partial}{\partial x}} \psi(x) = \frac{1}{2}\int_{x-t}^{x+t} \psi(\xi)\,d\xi.$$

Finally, for any $\varphi(x) \in L_2(\mathbb{R}^1)$, $\psi(x) \in L_2(\mathbb{R}^1)$ the formulae (4.24), (4.25) give the classical formula of d'Alembert

$$u(t,x) = \frac{\varphi(x+t) + \varphi(x-t)}{2} + \frac{1}{2}\int_{x-t}^{x+t} \psi(\xi)\,d\xi.$$

3. Cauchy problem for the Laplace equation. In the halfplane $\mathbb{R}^2_+ = \{ t > 0,\ x \in \mathbb{R}^1$ we consider the problem

$$\frac{\partial^2 u}{\partial t^2} + \frac{\partial^2 u}{\partial x^2} = 0, \tag{4.27}$$

$$u(0,x) = \varphi(x), \; \frac{\partial u}{\partial t}(0,x) = \psi(x). \tag{4.28}$$

In this case, the method, described in the previous examples, gives the formula

$$u(t,x) = \frac{e^{it\frac{\partial}{\partial x}} + e^{-it\frac{\partial}{\partial x}}}{2}\varphi(x) + \frac{e^{it\frac{\partial}{\partial x}} - e^{-it\frac{\partial}{\partial x}}}{2\frac{\partial}{\partial x}}\psi(x) \tag{4.29}$$

or, equivalently,

$$u(t,x) = \left[\cos t\frac{\partial}{\partial x}\right]\varphi(x) + \left[\frac{\sin t\frac{\partial}{\partial x}}{\frac{\partial}{\partial x}}\right]\psi(x). \tag{4.30}$$

As with formulae (4.24), (4.25), these formulae are correct in the sense of $L_2(\mathbb{R}^1)$ for any $t > 0$, if the functions $\varphi(x)$ and $\psi(x)$ belong to the space $W^\infty$. Thus, for any initial data $\varphi(x)$ and $\psi(x)$ from $W^\infty$ the Cauchy problem for the Laplace equation is correct, i. e. for any fixed $t > 0$, the change in $u(x,t)$ is small in the sense of $L_2(\mathbb{R}^1)$, if the change in the initial functions $\varphi(x)$ and $\psi(x)$ is small enough.

Taking into account, that any function $\varphi(x) \in W^\infty$ is an entire function, we can represent the formulae (4.29), (4.30) in the form

$$u(t,x) = \frac{\varphi(x+it) + \varphi(x-it)}{2} + \frac{1}{2i}\int_{x-it}^{x+it}\psi(\xi)\,d\xi. \tag{4.31}$$

(Let us note that it is impossible to give some correct reasonings about the closure of (4.31) up to any functions from $L_2(\mathbb{R}^1)$, since complex translations take place in the formula (4.31). In view of this fact, formula (4.31) is not valid for $\varphi(x) \in L_2(\mathbb{R}^1)$.)

4. Cauchy problem for the heat equation. In the halfplane $\mathbb{R}^2_+ = \left\{t > 0, \; x \in \mathbb{R}^1\right\}$ the heat equation

$$\frac{\partial u}{\partial t} - \frac{\partial^2 u}{\partial x^2} = 0 \tag{4.32}$$

under the initial condition

$$u(0,x) = \varphi(x) \tag{4.33}$$

is considered.

Putting $p = \partial^2/\partial x^2$, we have

$$u(t,x) = e^{tp}\, c(x),$$

where c(x) is an arbitrary function. From the condition (4.33) it follows that $c(x) = \varphi(x)$ and, consequently, the desired solution has the form

$$u(t,x) = e^{t\frac{\partial^2}{\partial x^2}}\varphi(x). \qquad (4.34)$$

It is obvious that the formula (4.34) is correct in $L_2(\mathbb{R}^1)$ for any $\varphi(x) \in W_1^\infty$, where

$$W_1^\infty = \left\{\varphi(x) \in C^\infty(\mathbb{R}^1):\ \sum_{n=0}^{\infty} \frac{t^n}{n!}\|D^{2n}\varphi(x)\|_2 < \infty \quad , \quad \forall t > 0\right\}$$

(the nontriviality of $W_1^\infty$ follows from Theorem 2.1, Ch. I). Let us show that the solution (4.34) is identically equal to the classical Poisson integral, i. e.

$$\exp\left(t\frac{\partial^2}{\partial x^2}\right)\varphi(x) = \frac{1}{2\sqrt{\pi t}}\int_{-\infty}^{\infty} e^{-\frac{(x-\xi)^2}{4t}}\varphi(\xi)\,d\xi.$$

Indeed, in view of convergence of the series

$$\sum_{n=0}^{\infty} \frac{t^n}{n!}\|D^{2n}u\|_2,\quad t > 0,$$

it follows that any function $\varphi(x) \in W_1^\infty$ is an entire function of a real variable. Consequently,

$$\frac{1}{2\sqrt{\pi t}}\int_{-\infty}^{\infty} e^{-\frac{(x-\xi)^2}{4t}}\varphi(\xi)\,d\xi = \frac{1}{2\sqrt{\pi t}}\int_{-\infty}^{\infty} e^{-\frac{(x-\xi)^2}{4t}}\cdot\sum_{m=0}^{\infty}\frac{D^m\varphi(x)}{m!}$$

$$\cdot(\xi - x)^m d\xi = \frac{1}{2\sqrt{\pi t}}\sum_{m=0}^{\infty}\frac{D^m\varphi(x)}{m!}\int_{-\infty}^{\infty} e^{-\frac{(x-\xi)^2}{4t}}(\xi - x)^m d\xi$$

$$= \sum_{n=0}^{\infty}\frac{4^n t^n D^{2n}\varphi(x)}{\sqrt{\pi}\,(2n)!}\int_{-\infty}^{\infty} e^{-\eta^2}\eta^{2n}d\eta$$

$$= \sum_{n=0}^{\infty}\frac{4^n t^n D^{2n}\varphi(x)}{\sqrt{\pi}\,(2n)!}\,\Gamma\left(n+\frac{1}{2}\right),$$

where

$$\Gamma(\alpha) = \int_0^{\infty} x^{\alpha-1}\,e^{-x}dx$$

is the Gamma-function of Euler. (Let us notice that in the process of these calculations the substitution $\xi = x + 2\sqrt{t}\,\eta$ was performed.)

Taking into account the known equality

$$\Gamma(n+\tfrac{1}{2}) = (n-\tfrac{1}{2})(n-\tfrac{3}{2})\ldots\tfrac{3}{2}\cdot\tfrac{1}{2}\,\Gamma(\tfrac{1}{2}) = \frac{(2n-1)!}{2^{2n-1}(n-1)!}\sqrt{\pi},$$

we get the desired formula

$$\frac{1}{2\sqrt{\pi t}}\int_{-\infty}^{\infty} e^{-\frac{(x-\xi)^2}{4t}}\varphi(\xi)\,d\xi \equiv \sum_{n=0}^{\infty}\frac{t^n}{n!}D^{2n}\varphi(x),$$

where $\varphi(x)\in W_1^\infty$ is an arbitrary function.

Since the Poisson integral is defined, obviously, for any function $\varphi(x)\in L_2(\mathbb{R}^1)$ and the imbedding $W_1^\infty\subset L_2(\mathbb{R}^1)$ is dense, we can regard (after the closure) that the formula (4.34) is defined for any $\varphi(x)\in L_2(\mathbb{R}^1)$ and gives the solution of the Cauchy problem (4.32), (4.33) for any function $\varphi(x)\in L_2(\mathbb{R}^1)$.

5. Cauchy problem for the inverse heat equation. In the halfplane $\mathbb{R}_+^2 = \left\{t>0,\ x\in\mathbb{R}^1\right\}$ we consider the problem

$$\frac{\partial u}{\partial t} + \frac{\partial^2 u}{\partial x^2} = 0,$$

$$u(0,x) = \varphi(x),\ x\in\mathbb{R}^1.$$

Putting $p = \partial^2/\partial x^2$, we find that

$$u(t,x) = e^{-tp}\,c(x),$$

where $c(x)$ is an arbitrary function. Obviously, $c(x)\equiv\varphi(x)$ and, consequently,

$$u(t,x) = e^{-t\frac{\partial^2}{\partial x^2}}\varphi(x).$$

For any $t>0$ the operator

$$e^{-t\frac{\partial^2}{\partial x^2}} \equiv \sum_{n=0}^{\infty}\frac{(-1)^n t^n}{n!}\frac{\partial^{2n}}{\partial x^{2n}}$$

is an elliptic differential operator of infinite order, the symbol of which is

$$a(\xi) \equiv \exp t\xi^2,\ \xi\in\mathbb{R}^1.$$

According to Theorem 4.1 the operator $e^{-t\frac{\partial^2}{\partial x^2}}$ gives an isomorphism

$$e^{-t\frac{\partial^2}{\partial x^2}}: H_a^\infty \longrightarrow \mathbb{R}^1,$$

where

$$H_a^\infty = \left\{ \varphi(x) \in L_2(\mathbb{R}^1) : \| e^{t\xi^2} \tilde{u}(\xi) \|_2 < \infty , \quad \forall t > 0 \right\}.$$

Thus, for any initial function $\varphi(x) \in H_a^\infty$ there exists one and only one solution of the Cauchy problem of the inverse heat equation in the sense of $L_2(\mathbb{R}^1)$.

After repeating the calculations, which were made in the previous example, we get that for any $\varphi(x) \in H_a^\infty$

$$e^{-t\frac{\partial^2}{\partial x^2}} \varphi(x) = \frac{1}{2\sqrt{\pi t}} \int_{-\infty}^{\infty} e^{\frac{(i\xi)^2}{4t}} \varphi(x + i\xi)\, d\xi , \quad t > 0,$$

i. e. for such initial functions the solution of the Cauchy problem for the inverse heat equation is given by the formula

$$u(t,x) = \frac{1}{2\sqrt{\pi t}} \int_{-\infty}^{\infty} e^{\frac{(i\xi)^2}{4t}} \varphi(x + i\xi)\, d\xi , \quad t > 0$$

(the existence of this complex convolution is verified with the Parseval equality).

6. Cauchy problem for the general homogeneous equation. In the halfplane $\mathbb{R}_+^2$ we study the following equation of order n with constant coefficients

$$\sum_{j=0}^{n} a_j \frac{\partial^n u(t,x)}{\partial t^j \partial x^{n-j}} = 0 \qquad (a_n \neq 0) \tag{4.35}$$

with Cauchy data

$$u(0,x) = \varphi_0(x), \ldots, \frac{\partial^{n-1} u(0,x)}{\partial t^{n-1}} = \varphi_{n-1}(x), \tag{4.36}$$

where $\varphi_0(x), \ldots, \varphi_{n-1}(x)$ are some given functions.

Let us put $p = i\partial/\partial x$ and solve the ordinary differential equation

$$\sum_{j=0}^{n} a_j (-ip)^{n-j} \frac{\partial^j u}{\partial t^j} = 0 \tag{4.37}$$

under conditions (4.36).

Obviously, the solution of the equation (4.37) may be represented in the form

$$u(t,x) = \sum_{m=0}^{n} u_m(t,p) \varphi_m(x),$$

where $u_m(t,p)$ are "basic" solutions of (4.37) under initial conditions

$$\frac{\partial^K}{\partial t^K} u_m(0,p) = \begin{cases} 0, & K \neq m \\ 1, & K = m. \end{cases}$$

Let us suppose (for brevity) that the characteristic polynomial of equation (4.37)

$$\mathcal{P}_n(\lambda,p) \equiv \sum_{j=0}^{n} a_j(-ip)^{n-j}\lambda^j$$

has the simple roots $\lambda_1,\dots,\lambda_n$. Since $\mathcal{P}_n(t,\lambda)$ is a homogeneous polynomial, these roots have the form

$$\lambda_K = \omega_K p, \quad K = 1,\dots,n,$$

where $\omega_K$ are the roots of the polynomial $\mathcal{P}(\lambda,1)$.

An elementary calculation shows that the basic solutions are

$$u_m(t,p) = \sum_{K=1}^{n} e^{\omega_K tp} \frac{V_{m+1,K}}{V\, p^m},$$

where $V = V\left[\omega_1,\dots,\omega_n\right]$ is Vandermonde's determinant and $V_{m+1,K}$ are its algebraic minors.

Since

$$\sum_{K=1}^{n} \omega_K^s V_{m+1,K} = 0, \quad s = 0,1,\dots,m-1,$$

then the function $u_m(t,p)$ is an entire function of p, i. e. the function $u_m(t,p)$ defines the differential operator of infinite order

$$u_m(t,i\tfrac{\partial}{\partial x}) \equiv \frac{1}{V} \sum_{K=1}^{n} \frac{V_{m+1,K}}{(i\frac{\partial}{\partial x})^m} \exp\,(it\omega_K \tfrac{\partial}{\partial x})$$

(the coefficients of this operator decrease as factorial-type coefficients). Therefore, the solution of the problem (4.35), (4.36) is defined in the space $W^\infty$ by the formula

$$u(t,x) = \sum_{m=0}^{n-1} u_m(t,i\tfrac{\partial}{\partial x})\varphi_m(x). \tag{4.38}$$

Since any function from the space $W^\infty$ is an entire function, then for $\varphi(x) \in W^\infty$

$$\exp(it\omega_K \tfrac{\partial}{\partial x})\varphi(x) = \varphi(x + it\omega_K),$$

and, therefore, the formula (4.38) may be written in the form

$$u(t,x) = \frac{1}{V}\sum_{K=1}^{n} V_{1,K}\,\varphi_0(x + it\omega_K)$$

$$+ \frac{1}{V}\sum_{m=1}^{n-1}\frac{1}{(m-1)!}\sum_{K=1}^{n} V_{m+1,K}\int_0^{x+it\omega_K}(x + it\omega_K)^{m-1}\,\varphi_m(\xi)\,d\xi. \quad (4.39)$$

Let us notice one special case of the latter formula, namely, the case $\omega_K = i\alpha_K$, $\alpha_K \in \mathbb{R}^1$, $K = 1,\dots,n$. It follows that the initial equation (4.35) is a strongly hyperbolic equation. In this case all translations in (4.39) are real and (in view of the density of the imbedding $W^\infty \subset L_2(\mathbb{R}^n)$) this formula is valid for all $\varphi_m(x) \in L_2(\mathbb{R}^n)$. Thus, in the case of a strongly hyperbolic equation the Cauchy problem is correct in the classical sense in $L_2(\mathbb{R}^1)$.

On the other hand, if only one translation in (4.39) is complex, then this formula does not admit a closure up to the functions $\varphi_m(x)$ having a finite smoothness. Therefore, the strongly hyperbolic equation is the unique equation of the type (4.35) for which the Cauchy problem is correct in the classical sense.

7. Dirichlet problem for an elliptic equation in the cylinder. Let $y \in (-\frac{1}{2},\frac{1}{2})$, $x \in G$, where $G \subset \mathbb{R}^\nu$ is a domain with the boundary $\Gamma$. In the cylinder $Q = \left[-\frac{1}{2},\frac{1}{2}\right] \times G$ the Dirichlet problem

$$-\frac{\partial^2 u}{\partial y^2} + (-1)^m \Delta^m u = 0, \quad y \in (-\tfrac{1}{2},\tfrac{1}{2}), \quad x \in G, \quad (4.40)$$

$$u(\pm\tfrac{1}{2},x) = \varphi_\pm(x), \quad x \in G, \quad (4.41)$$

$$D^\omega u\big|_S = 0, \quad |\omega| \leq m-1 \quad (4.42)$$

is considered. Here $\Delta^m$ is a polyharmonic operator, $\varphi_\pm(x)$ are given functions, $S = (-\frac{1}{2},\frac{1}{2}) \times \Gamma$ is the lateral surface of the cylinder Q.

To solve this problem let us put $p^2 = (-\Delta)^m$ and consider the ordinary differential equation

$$-\frac{\partial^2 u}{\partial y^2} + p^2 u = 0.$$

We have

$$u(y,x) = e^{yp}\,c_1(x) + e^{-yp}\,c_2(x),$$

where $c_1(x)$ and $c_2(x)$ are arbitrary functions.

Defining $c_1(x)$ and $c_2(x)$ from the boundary conditions (4.41), we get the formula

$$u(y,x) = \frac{\operatorname{sh}(y+\frac{1}{2})p}{p}\, u_+(x) - \frac{\operatorname{sh}(y-\frac{1}{2})p}{p}\, u_-(x), \tag{4.43}$$

where the functions $u_{\pm}(x)$ satisfy the equations

$$\frac{\operatorname{sh} p}{p}\, u_{\pm}(x) = \varphi_{\pm}(x). \tag{4.44}$$

In order to obtain the nonformal solution from the formula (4.43), (4.44) let us notice that any smooth solution $u(y,x)$ of our initial problem satisfies conditions

$$D^{\omega}\Delta^{mn}u\big|_{\Gamma} = 0,\ n = 0,1,\dots,\ y \in (-\tfrac{1}{2},\tfrac{1}{2})$$

(this follows immediately from the equation (4.40) and the boundary conditions (4.42)).

Consequently, in conformity with (4.44) the functions $u_{\pm}(x)$ must be the solutions of the boundary problem of infinite order

$$\frac{\operatorname{sh}(-\Delta)^{\frac{m}{2}}}{(\Delta)^{\frac{m}{2}}}\, u_{\pm}(x) \equiv \sum_{n=0}^{\infty} \frac{(-1)^{mn}\Delta^{mn}}{2(2n+1)!}\, u_{\pm}(x) = \varphi_{\pm}(x),\ x \in G, \tag{4.45}$$

$$D^{\omega}\Delta^{mn}u_{\pm}(x)\big|_{\Gamma} = 0,\ n = 0,1,\dots \tag{4.46}$$

Let $W^{\infty}\left\{\frac{1}{(2n+1)!}, \nabla^{m}\right\}$ be the energy space of the problem (4.45), (4.46), i. e.

$$W^{\infty}\left\{\frac{1}{(2n+1)!}, \nabla^{m}\right\} = \left\{ u(x) \in C^{\infty}(G):\ \sum_{n=0}^{\infty} \frac{1}{(2m+1)!} \|\nabla^{mn}u\|_2^2 < \infty\ ,\right.$$

$$\left. D^{\omega}\Delta^{mn}u\big|_{\Gamma} = 0,\ n = 0,1,\dots\right\}.$$

Then, according to Ch. II, for any $\varphi_{\pm}(x) \in W^{-\infty}\left\{\frac{1}{(2n+1)!}, \nabla^{m}\right\}$ there exists one and only one solution $u_{\pm}(x) \in W^{\infty}\left\{\frac{1}{(2n+1)!}, \nabla^{m}\right\}$ of the problem (4.45), (4.46). In this case the formula (4.43) gives the solution of the initial Dirichlet problem such that for any $y \in (-\frac{1}{2},\frac{1}{2})$ it is the generalized solution (with its derivatives) on the basic space $W^{\infty}\left\{\frac{1}{(2n+1)!}, \nabla^{m}\right\}$.

If the functions $\varphi_{\pm}(x) \in L_2(G)$, then the formula (4.43) gives the classical solution.

Indeed, as is known, the Dirichlet problem

$$(-1)^m\Delta^m u = 0, \quad D^{\omega}u\big|_{\Gamma} = 0, \quad |\omega| \leq m-1,$$

has a system of eigenfunctions $v_1(x), v_2(x), \ldots$, which form an orthogonal basis in $L_2(G)$. Without loss of generality one can assume that this basis is an orthonormal basis. Therefore, any function $u(x) \in L_2(G)$ can be expanded in the series

$$u(x) = \sum_{n=1}^{\infty} u_k v_k(x),$$

where $u_k = (u, v_k)$.

Let us denote by $W^{\infty}\left\{\frac{1}{(2n+1)!}, \Delta^m\right\}$ the space of functions $u(x) \in C^{\infty}(G)$, which satisfy the conditions (4.46) and have the finite norm

$$\|u\|_{\infty}^2 = \sum_{k=1}^{\infty} u_k^2 \left(\sum_{n=0}^{\infty} \frac{\lambda_k^n}{(2n+1)!}\right)^2,$$

where $\lambda_k > 0$ are the eigenvalues, which correspond to the eigenfunctions $v_k(x)$, $k = 1, 2, \ldots$ (Let us notice that the nontriviality of the space $W^{\infty}\left\{\frac{1}{(2n+1)!}, \Delta^m\right\}$ is evident, since all finite linear combinations of $v_1(x), v_2(x), \ldots$ belong to this space.)

It is easy to see (cf. with p. II of present paragraph) that for any $\varphi_{\pm}(x) \in L_2(G)$ there exists only one solution $u_{\pm}(x) \in W^{\infty}\left\{\frac{1}{(2n+1)!}, \Delta^m\right\}$. The formula (4.43) defines in this case the classical solution $u(x) \in L_2(G)$ and such that

$$\frac{\partial^2 u}{\partial y^2} \in L_2(G), \quad \Delta^m u \in L_2(G)$$

too.

8. Cauchy problem for one differential equation of relativistic quantum mechanics. In the halfplane $\mathbb{R}_+^2$ we consider the problem (see J. BJORKEN, S. DRELL [1], Ch. I, § 1)

$$i\frac{\partial u}{\partial t} - \sqrt{I - \Delta}\, u = 0, \quad t > 0, \ x \in \mathbb{R}^1, \tag{4.47}$$

$$u(0,x) = \varphi(x), \quad x \in \mathbb{R}^1. \tag{4.48}$$

We shall consider the operator $\sqrt{I - \Delta}$ as the differential operator of infinite order

$$\sqrt{I - \Delta} \equiv \sum_{n=0}^{\infty} C_{1/2}^n (-\Delta)^n$$

in the space $L_2(\mathbb{R}^1)$ with domain of definition

$$H^{\infty} = \left\{ u(x) \in L_2(\mathbb{R}^1) : \sum_{n=0}^{\infty} C_{1/2}^{n}(-\Delta)^n u(x) \in L_2(\mathbb{R}^1) \right\}.$$

The simple calculations show that for $|\xi| > 1$

$$a_N^2(\xi) \equiv \left( \sum_{n=0}^{N} C_{1/2}^{n}\ \xi^{2n} \right)^2 \longrightarrow +\infty$$

as $N \to \infty$. From this it follows that $\tilde{u}(\xi) \equiv 0$ for $|\xi| > 1$, i. e.

$$H^{\infty} = \left\{ u(x) \in L_2(\mathbb{R}^1) : \tilde{u}(\xi) \equiv 0,\ |\xi| > 1 \right\}.$$

The norm in $H^{\infty}$ may be taken as the norm $\|u(x)\|_2$ in $L_2(\mathbb{R}^1)$ or

$$\left\| \sqrt{I - \Delta}\ u(x) \right\|_2 = \left\| \sqrt{1 + \xi^2}\ \tilde{u}(\xi) \right\|_2 .$$

It follows that the operator $\sqrt{I - \Delta}$ is a bounded operator in $H^{\infty}$. Consequently,

$$u(t,x) = e^{-it\sqrt{I-\Delta}}\varphi(x), \quad \varphi(x) \in H^{\infty},$$

is the desired solution of the problem (4.47), (4.48). In terms of Fourier transforms,

$$\tilde{u}(t,\xi) = \begin{cases} 0, & |\xi| > 1 \\ \exp(-it\sqrt{1 + \xi^2})\tilde{\varphi}(\xi), & |\xi| \leq 1. \end{cases}$$

Resume. Summing up all that was said in this paragraph, we get, obviously, a new method of solving some linear problems for partial differential equations (in particular, for classical problems of mathematical physics).

The foundation of this method is a nonformal algebra of differential operators of infinite order, which are generated by entire functions. The nonformality of this algebra is achieved by the consideration of differential operators of infinite order in the corresponding Sobolev spaces of infinite order. It permits us to introduce a parameter p (in a suitable way) and to solve the partial differential equation as an ordinary differential equation with some initial and boundary data. Finally, finding the solution of the initial problem involves either a direct application of differential operators of infinite order to the data of the problem or the solution of some problems for differential equations of infinite order, the dimension of which is less than the dimension of the initial problem. In the case of two variables the latter problems are problems for ordinary differential equations of infinite order.

It is necessary to say that the type of the differential equation of finite order plays no role if this equation is considered in the spaces $W^\infty$. However, we emphasize that for those problems which are correct in spaces of finite smoothness, both the spaces $W^\infty$ themselves and the differential problems of infinite order play an intermediate role and are the means of the investigation of the initial problem. At the same time, the introduction of the spaces $W^\infty$ for those problems which are noncorrect in the usual sense, is the core of this method: problems which are noncorrect in the classical sense are correct in these spaces.

Let us also remark that in view of the general principle of duality this method may be applied to solve the same partial differential equations in the sense of distributions on the basic space $W^\infty$.

In conclusion, we notice that the method described here may be applied to integro-differential equations and, in general, to differential-operator equations of the type

$$\sum_{j=0}^{n} A_j \frac{d^j u(t)}{dt^j} = h(t), \qquad (4.49)$$

where each operator $A_j \equiv a_j A^{m_j}$, $a_j \in \mathbb{C}^1$, $m_j \in \mathbb{N}$; and A is an operator (in general, unbounded) in a Banach or Hilbert space. In this case we put p = A and solve the corresponding problem for an ordinary differential equation

$$\sum_{j=0}^{n} a_j p^{m_j} \frac{d^j u}{dt^j} = h(t),$$

where p is a parameter. The answer obtained after calculations will be nonformal, if the operators of the type exp(tA), cos(tA) and others which arise during these calculations are considered in the spaces

$$W^\infty = \left\{ u \in \bigcap_{n=0}^{\infty} D(A^n) : \sum_{n=0}^{\infty} a_n |A^n u| < \infty \right\},$$

where $a_n \geqq 0$ are the corresponding coefficients.

Corresponding generalizations for the case of many variables and for the case of the differential-operator equation (4.49), where

$$A_j = a_j A_1^{m_{j1}} \dots A_n^{m_{jn}},$$

introduce no principal difficulties.

## § 5. A stochastic problem of the theory of elasticity

Let $G \subset \mathbb{R}^3$ be a region (an elastic body) on which an external force $F(x) = \{F_1(x), F_2(x), F_3(x)\}$, $x \in G$, acts.

Let the tensor of elasticity $c_{ijpq}(x)$, the tensor of intensity $\sigma_{ij}(x)$ and the tensor of deformations $\varepsilon_{ij}(x)$ be random functions which are determined on a certain probability space and which depend on $x \in G$ as a parameter. Exterior forces $F_i(x)$, $i = 1,2,3$, will be assumed to be determined forces.

It is known that the basic equations of linear elasticity theory have the following form

$$\partial_j \sigma_{ij}(x) + F_i(x) = 0, \tag{5.1}$$

$$\varepsilon_{ij}(x) = \frac{1}{2}(\partial_j u_i(x) + \partial_i u_j(x)), \quad i = 1,2,3, \tag{5.2}$$

where $u(x) = \{u_1(x), u_2(x), u_3(x)\}$ is the corresponding vector of elastic dislocations. Moreover, let intensities and deformations be connected by Hooke's law

$$\sigma_{ij}(x) = c_{ijpq}(x) \cdot \varepsilon_{pq}(x) \tag{5.3}$$

(Here, as usual, there is a summation for the same indices.)

In the case of a deterministic problem the substitution of (5.3) and (5.2) into (5.1) gives a system of three equations for $u(x) = \{u_1(x), u_2(x), u_3(x)\}$. One can solve this system with suitable boundary conditions.

Under our stochastic conditions this method gives a system of three equations with random coefficients. In this case the problem of the definition of mathematical expectation of random values $\sigma_{ij}(x)$, $\varepsilon_{ij}(x)$, $u_i(x)$ naturally arises.

Let $\langle \sigma_{ij}(x) \rangle$, $\langle \varepsilon_{ij}(x) \rangle$, $\langle u_i(x) \rangle$ be the mean values of $\sigma_{ij}(x)$, $\varepsilon_{ij}(x)$, $u_i(x)$ along the set of realizations (in the stationary case these mean values do not depend on $x \in G$ and equal the mean values in volume).

Taking the mean values of (5.1) - (5.3) we obtain the following system

$$\partial_j \langle \sigma_{ij}(x) \rangle + F_i(x) = 0, \tag{5.4}$$

$$\langle \varepsilon_{ij}(x) \rangle = \frac{1}{2}(\partial_i \langle u_j(x) \rangle + \partial_j \langle u_i(x) \rangle), \tag{5.5}$$

$$\langle \sigma_{ij}(x)\rangle = \langle c_{ijpq}(x)\varepsilon_{pq}(x)\rangle . \tag{5.6}$$

In the general case there is no simple law between the mean values $\langle c_{ijpq}(x)\varepsilon_{pq}(x)\rangle$ and $\langle c_{ijpq}(x)\rangle$ , $\langle \varepsilon_{pq}(x)\rangle$ , and it is one of the major difficulties of stochastic mechanics.

V. V. NOVOŠILOV [1] obtained such a law in the case of micro-non-homogeneous statistical isotopic bodies. Moreover, in the case of stationary random functions $c_{ijpq}(x)$ [1)] this law has the following form

$$\langle \sigma_{ij}(x)\rangle = \sum_{k=0}^{\infty} \sum_{|\alpha|=2k} (-1)^k c_{\alpha ijpq} D^{\alpha} \langle \varepsilon_{pq}(x)\rangle , \tag{*}$$

where the constant tensors $c_{\alpha ijpq}$ are calculated with the aid of tensors $c_{ijpq}(x)$. In general, there is such a procedure in the work of V. M. LEVIN [1].

<u>Remark</u>. M. V. PAUKŠTO [1] proved that in the case $G = \mathbb{R}^3$ the relation (*) is equivalent (under some additional conditions) to Hooke's law for a homogeneous medium with a bounded long-influence.

Putting for $|\alpha| = 2k$, $|\beta| = |\gamma| = k$

$$r_{\alpha} = \sum_{\beta+\gamma=\alpha} 1, \quad c^{o}_{\beta\gamma ijpq} = c_{\beta+\gamma,ijpq}\, r^{-1}_{\beta+\gamma},$$

we may write the relation (*) in the form

$$\langle \sigma_{ij}(x)\rangle = \sum_{|\beta|=|\gamma|=0}^{\infty} (-1)^{|\beta|} c^{o}_{\beta\gamma ijpq} D^{\beta+\gamma} \langle \varepsilon_{pq}(x)\rangle .$$

Substituting this formula in (5.4) and taking into account (5.5), we obtain the system of partial differential equations of infinite order

$$\sum_{|\alpha|=|\beta|=1}^{\infty} (-1)^{|\alpha|} D^{\alpha}(a_{\alpha\beta} D^{\beta} \langle u(x)\rangle ) = F(x), \quad x \in G, \tag{5.7}$$

where the elements $a_{\alpha\beta ij}$ $(i,j = 1,2,3)$ of matrices $a_{\alpha\beta}$ are determined by formula

$$a_{\alpha\beta ij} = \sum_{\substack{\alpha'+e_p=\alpha \\ \beta'+e_q=\beta}} \frac{1}{2} (c^{o}_{\alpha'\beta' ipqj} + c^{o}_{\alpha'\beta' ipjq})$$

---

1) Even though the $c_{ijpq}(x)$ are stationary random functions, the tensors $\sigma_{ij}(x)$ and $\varepsilon_{ij}(x)$ will be in general some nonstationary functions in view of the micro-non-homogeneousness of the body (even for determinate exterior forces $F_i(x)$).

($e_p$ is a multiindex of type $(\underbrace{0,\dots,0}_{p-1},1,0,\dots,0))$.

Let the following conditions be fulfilled:

1) For every $m \geqq 1$ and any set of $f_\alpha$, $|\alpha| = m$, there is the inequality

$$\sum_{|\alpha|=|\beta|=m} a_{\alpha\beta} f_\alpha f_\beta \geqq b_m \sum_{|\alpha|=m} f_\alpha f_\alpha,$$

where $b_m > 0$ is a number sequence;

2) $|a_{\alpha\beta}| = O(b_m)$ for $|\alpha| = |\beta| = m$, $m \to \infty$.

Under these assumptions the system (5.7) is an elliptic coercive system of infinite order, where $\langle u_1(x)\rangle$, $\langle u_2(x)\rangle$, $\langle u_3(x)\rangle$ are the unknown functions. Therefore, the above theory of elliptic systems of infinite order may be applied to this system; namely, this system may be solved with boundary conditions such that the corresponding spaces of infinite order are nontrivial.

It is possible, for example, to consider the following conditions:

I. Dirichlet conditions of infinite order

$$D^\omega u|_\Gamma = 0, \quad |\omega| = 0,1,\dots; \tag{5.8}$$

II. periodic conditions

$$D^\omega u(x + 2\pi) = D^\omega u(x), \quad |\omega| = 0,1,\dots \tag{5.9}$$

etc. (we do not consider the question of mechanical realization of these conditions).

Let us consider these problems.

I. Dirichlet problem of infinite order. Taking into account the above assumptions and the results of Chapter II we obtain that the "energy" space of the problem (5.8), (5.7) is

$$\mathring{W}^\infty\{b_m, 2\} = \left\{u(x) \in C_0^\infty(G) : \varrho(u) \equiv \sum_{m=1}^\infty b_m \sum_{|\alpha|=m} \|D^\alpha u\|_2^2 < \infty\right\}.$$

(There is no $m = 0$ in the formula which defines the metric $\varrho(u)$, but it plays no role because the functions $u(x)$ equal zero on the boundary.)

As a corollary of the results of the present chapter we obtain the following.

Theorem 5.1. Let $\overset{\circ}{W}{}^{\infty}\{b_m, 2\}$ be nontrivial. Then for any right side $F(x) \in W^{-\infty}\{b_m, 2\}$ there exists one and only one solution $u(x) \in \overset{\circ}{W}{}^{\infty}\{b_m, 2\}$ of the Dirichlet problem (5.7), (5.8). Moreover, the Galerkin approximations tend to u(x) uniformly (local uniformly) with all derivatives.

II. Periodic problem. The analogous space of periodic functions

$$W^{\infty}\{b_m, 2\} = \left\{u(x) \in C^{\infty}(T^n) : \varrho(u) \equiv \sum_{m=1}^{\infty} b_m \sum_{|\alpha|=m} \|D^{\alpha}u\|_2^2 < \infty\right\}$$

corresponds to this problem.

Let us remark, however, that in this case the norm $\varrho(u)$ has a nontrivial kernel; namely, if u(x) = const, then $\varrho(u) = 0$. Therefore, the previous theory cannot be applied formally. This shortcoming is eliminated if we consider only those functions u(x) such that u(x) is orthogonal to unity, i. e. (u,1) = 0.

In other words, let $V^{\infty}\{b_m, 2\}$ be the subspace of the space $W^{\infty}\{b_m, 2\}$ which contains precisely the periodic functions u(x) having a Fourier expansion of the following form

$$u(x) = \sum_{|\alpha|=1}^{\infty} c_{\alpha} \exp\{ix\alpha\}$$

(there is no constant in this series). Then the following theorem holds.

Theorem 5.2. Let the space $V^{\infty}\{b_m, 2\}$ be nontrivial. Then for any function $h(x) \in V^{-\infty}\{b_m, 2\}$ the periodic problem (5.7), (5.9) has one and only one solution $u(x) \in V^{\infty}\{b_m, 2\}$. Moreover, the Galerkin approximates tend to u(x) uniformly with all its derivatives.

In particular, Theorem 5.2 means that for $F(x) \in L_2(T^n)$ equation (5.7) is solvable if and only if F(x) is orthogonal to unity.

Remark. Let us note that the question of the study of the stochastic problem, described here, by means of a direct investigation of system (5.7) was posed in the works of V. V. NOVOŠILOV [1] and V. M. LEVIN [1].

## CHAPTER III

# TRACE THEORY AND INHOMOGENEOUS DIRICHLET PROBLEM OF INFINITE ORDER

## Introduction

One of the aims of the present Chapter is to study the problem

$$\sum_{|\alpha|=0}^{\infty} (-1)^{|\alpha|} D^{\alpha} A_{\alpha}(x, D^{\gamma} u) = h(x), \quad x \in G,$$

$$D^{\omega} u\big|_{\Gamma} = f_{\omega}(x'), \quad |\omega| = 0, 1, \ldots, \quad x' \in \Gamma,$$

i. e. to solve the Dirichlet problem of infinite order with inhomogeneous boundary conditions.

It turns out that (as in the case of equations of finite order) in going over from the homogeneous boundary conditions

$$D^{\omega} u\big|_{\Gamma} = 0, \quad |\omega| = 0, 1, \ldots,$$

to the inhomogeneous conditions one does not have principal difficulties as long as we have a suitable method for solving the homogeneous problem and the boundary values $f_{\omega}(x')$, $|\omega| = 0, 1, \ldots$, admit an extension into the domain G as a function $u(x) \in W^{\infty}\{a_{\alpha}, p_{\alpha}\}$ so that $D^{\omega} u|_{\Gamma} = f_{\omega}(x')$, $|\omega| = 0, 1, \ldots$ Therefore, the inhomogeneous boundary value problem will be studied, if we construct the theory of traces of functions from the Sobolev spaces of infinite order $W^{\infty}\{a_{\alpha}, p_{\alpha}\}$.

§§ 1, 2 of the present Chapter are devoted to this question. Namely, in § 1 a criterion of the trace for the case of an arbitrary domain G is obtained. From our point of view this criterion, being universal, is not easily verified. It, therefore, seems natural to study, along with the trace criterion, the question of sufficient but easily verifiable conditions on $f_{\omega}(x')$, $|\omega| = 0, 1, \ldots$, under which there is an extension to $W^{\infty}\{a_{\alpha}, p_{\alpha}\}$.

Here this question is solved for the case of a strip in $\mathbb{R}^{n+1}$. Further results in this direction are discussed. Finally, after obtaining "trace" results we prove the correctness of the inhomogeneous Dirichlet problem of infinite order.

## § 1. Trace criterion

Let $G \subset \mathbb{R}^n$ be an arbitrary domain the boundary of which is denoted as $\Gamma$. Let us consider in this domain the space

$$W^\infty\{a_\alpha, p_\alpha\} = \left\{u(x) \in C^\infty(G) : \varrho(u) \equiv \sum_{|\alpha|=0}^{\infty} a_\alpha \|D^\alpha u\|_p^p < \infty\right\},$$

where $a_\alpha \geqq 0$, $p_\alpha \geqq 1$ are some constants, and let us pose the following question: What properties must the functions $f_\omega(x')$, $x' \in \Gamma$, $|\omega| = 0,1,\dots$, posses to imply the existence of a function $u(x) \in C^\infty(G)$ such that $D^\omega u|_\Gamma = f_\omega(x')$, $|\omega| = 0,1,\dots$, and, in addition, $\varrho(u) < \infty$ ? We will say in this case that the functions $f_\omega(x')$, $|\omega| = 0,1,\dots$, are the trace on $\Gamma$ of function $u(x) \in W^\infty\{a_\alpha, p_\alpha\}$ and $u(x)$ itself is an extension of a trace in this space.

Before formulating the main results let us make some observations. Namely, let $N > 0$ be a natural number. Let us suppose that for any $N > 0$ the boundary conditions $f_\omega(x')$, $|\omega| = 0,1,\dots,N-1$, can be extended into the interior of the domain $G$ as a function in the space

$$W_N = \bigcap_{|\alpha| \leqq N} W^\alpha_{p_\alpha}(G),$$

i. e. there exists a function $u(x) \in W_N$ such that

$$D^\omega u|_\Gamma = f_\omega(x'), \quad |\omega| \leqq N-1, \tag{1.1}$$

$$\varrho_N(u) \equiv \sum_{|\alpha|=0}^{N} a_\alpha \|D^\alpha u\|_p^p < \infty . \tag{1.2}$$

We denote the set all such extensions by $E_N$. It is evident that $E_N$ is a set of type

$$E_N = \{u(x) : f(x) + z(x)\},$$

where $f(x)$ is a function satisfying (1.1) and (1.2), i. e. an extension of boundary conditions (1.1), and $z(x)$ is an arbitrary function such that

$$D^\omega z|_\Gamma = 0, \quad |\omega| \leqq N-1; \quad \varrho(z) < \infty .$$

It follows that there exists a unique function $u_N(x) \in E_N$ such that

$$\varrho_N(u_N) = \inf \varrho_N(u), \quad u \in E_N.$$

Theorem 1.1. A family of boundary functions $f_\omega(x')$, $|\omega| = 0,1,\dots$, $x' \in \Gamma$, is a trace of a function $u(x) \in W^\infty\{a_\alpha, p_\alpha\}$ if and only if the following conditions are fulfilled:

1) For every $N = 1,2,\ldots$ the functions $f_\omega(x')$, $|\omega| \leq N-1$, $x' \in \Gamma$, admit an extension in the space $W_N$;

2) There exists a constant $K > 0$ such that for all N the inequality

$$\varrho_N(u_N) \leq K$$

is valid.

Proof. Indeed, if the family $f_\omega(x')$, $|\omega| = 0,1,\ldots$, $x' \in G$, is a trace of a function $u(x) \in W^\infty\{a_\alpha, p_\alpha\}$, then for any $N = 1,2,\ldots$ the family $f_\omega(x')$, $|\omega| = 0,1,\ldots,N-1$, $x' \in \Gamma$, is a trace of the same function in the space $W_N$. Moreover, it is clear that

$$\varrho_N(u_N) \leq \varrho_N(u) \leq \varrho(u) < \infty .$$

Conversely, if conditions 1) and 2) of our theorem are fulfilled, then, using the compactness of the imbedding $W_N \subset W_{N-1}$ and the diagonal process, we get that there exists a subsequence of the sequence $u_N(x)$ and a function $u(x) \in C^\infty(G)$ such that this subsequence converges to $u(x)$ uniformly (locally) with all its derivatives. It is clear that $u(x)$ satisfies the conditions

$$D^\omega u|_\Gamma = f_\omega(x'), \quad |\omega| = 0,1,\ldots, \quad x' \in \Gamma .$$

Moreover, from condition 2) we obtain that $\varrho(u) \leq K$, and hence $u(x)$ is the desired extension. The theorem is proved.

Remark. It is evident that in the case $p_\alpha > 1$ the functions $u_N(x)$ are the solutions of the problems (Euler equations of functional $\varrho_N(u)$)

$$\sum_{|\alpha| \leq N} (-1)^{|\alpha|} D^\alpha\left[(p_\alpha - 1) a_\alpha |D^\alpha u|^{p_\alpha - 2} D^\alpha u\right] = 0, \tag{1.3}$$

$$D^\omega u|_\Gamma = f_\omega(x'), \quad |\omega| \leq N-1, \tag{1.4}$$

and, consequently, the effectiveness of the trace criterion obtained in Theorem 1.1 directly depends on the effectiveness of the methods for solving the nonlinear problems (1.3), (1.4) and of estimates of their solutions. Both questions are difficult.

In connection with this remark we pay our attention to the case $p_\alpha \equiv 2$, i. e. the case of Hilbert space $W^\infty\{a_\alpha, 2\}$ in which the problems (1.3), (1.4) are linear problems.

Here our criterion can be formulated in algebraic form. In fact, from the linear elliptic theory it follows that in the case $p_\alpha \equiv 2$ the functions $u_N(x)$ are infinitely smooth. Multiplying (1.3) (where $p_\alpha \equiv 2$) by $u_N(x)$ and integrating by part we obtain that

$$\varrho_N(u_N) + \sum_{\substack{|\alpha| \leq N \\ \gamma+e_k \leq \alpha}} (-1)^{|\alpha|+|\gamma|} \int_\gamma D^{\alpha-\gamma-e_k}(a_\alpha D^\alpha u_N) D^\gamma u_N \cos(n,x_k)\, d\Gamma = 0, \tag{1.5}$$

where $e_k$ is the multiindex, corresponding to the derivative $\partial/\partial x_k$, n is the exterior normal on the boundary $\Gamma$. Moreover, the notation $\sum_{\gamma+e_k \leq \alpha}$ denotes the summation for all indices $\gamma+e_k \leq \alpha$ which successively arise in the process of integration by part for $x_1, x_2 \ldots$ etc.

Since the solution $u_N(x)$ and all its derivatives depend linearly on the $f_\omega(x')$, $|\omega| \leq N-1$, $x' \in \Gamma$, the relation (1.5) can be rewritten in the form

$$\varrho_N(u_N) + A_N(f_\omega, f_\omega) = 0,$$

where $A_N(f_\omega, f_\omega)$, $|\omega| \leq N-1$, is a quadratic form.

It is clear that $\varrho_N(u_N) \leq K$ if and only if the sequence of quadratic forms $A_N(f_\omega, f_\omega)$ is bounded as $N \to \infty$. Thus, in the linear case the verification of trace criterion is equivalent to the existence of the uniform estimates for the Green's functions of problems (1.3), (1.4).

## § 2. Sufficient algebraic conditions in the case of the strip

In this paragraph we give simple sufficient conditions on functions $\varphi_m(x)$ and $\psi_m(x)$ (see (2.1)) such that these functions are the trace of a function $u(t,x) \in W^\infty\{a_{n\alpha}, p_\alpha\}$ in the strip.

Let G be a strip in the space $\mathbb{R}^{\nu+1}$ of variables $(t,x)$, i. e. $G = [0,a] \times \mathbb{R}^{\nu+1}$ $(\nu \geq 1)$, where $t \in [0,a]$, $x \in \mathbb{R}$. In this strip we consider the nontrivial space (see § 4, Chapter I).

$$\mathring{W}^\infty\{a_{n\alpha}, p\} = \left\{ u(t,x) \in C^\infty(G) : \varrho(u) \equiv \sum_{n+|\alpha|=0}^{\infty} a_{n\alpha} \| D_t^n D_x^\alpha u \|_p^p < \infty, \right.$$

$$D_t^m u(0,x) = 0, \ D_t^m u(a,x) = 0, \ m = 0,1,\ldots ,$$

where $a_{n\alpha} \geq 0$, $p \geq 1$. It can be assumed without loss of generality that $a_{00} > 0$.

Futher, suppose two families of functions $\varphi_m(x)$ and $\psi_m(x)$, $m = 0,1,\dots,$ are given.

We want to find a function $u(t,x)$ such that

$$D_t^m u(0,x) = \varphi_m(x),\ D_t^m u(a,x) = \psi_m(x),\ m = 0,1,\dots \qquad (2.1)$$

As will be seen below, it is sufficient to consider the case $\psi_m(x) = 0$, $m = 0,1,\dots$, i. e. to consider such conditions on functions $\varphi_m(x)$ under which there exists a function $u(t,x) \in W^\infty\{a_{n\alpha},p\}$ satisfying the conditions

$$D_t^m u(0,x) = \varphi_m(x),\ D_t^m u(a,x) = 0,\ m = 0,1,\dots \qquad (2.2)$$

In order to formulate the main result, let us introduce the following numerical sequence

$$s_m = \sum_{k=0}^{\infty} \frac{M_{m+k}^c}{M_{m+k+1}^c}, \qquad (2.3)$$

where $M_n^c$ is a convex regularization by means of logarithms of the sequence

$$M_n = \begin{cases} b_n^{-\frac{1}{p}} & \text{if } b_n > 0, \\ +\infty & \text{if } b_n = 0 \end{cases}$$

($b_n$ are the numbers, which were defined in Theorem 4.1, Chapter I). As was already noted in § 1, Chapter I, the non-quasianalyticity of the Hadamard class $C\{M_n\}$ means that

$$\sum_{n=0}^{\infty} \frac{M_n^c}{M_{n+1}^c} < \infty, \qquad (2.4)$$

which implies that $s_m \to 0$ as $m \to \infty$.

Theorem 2.1. Let the following conditions be fulfilled:

1) the numbers

$$\varrho_m = \sup_n (M_n^c \sum_{|\alpha|=0}^{\infty} a_{n\alpha}^{1/p} |D^\alpha \varphi_m(x)|_p),$$

where $|\cdot|_p$ is the norm in $L_p(\mathbb{R}^\nu)$, are finite for $m = 0,1,\dots;$

2) there exists a number $r > 0$ such that

$$\sum_{m=0}^{\infty} \varrho_m \max\left[\frac{s_m r^m}{m!},\ (M_m^c)^{-1}\right] < \infty,$$

where $s_m$ are defined by formula (2.3).

Then there exists a function $u(t,x) \in W^{\infty}\{a_{n\alpha}, p\}$ satisfying conditions (2.2).

Proof. For the proof of the theorem the desired function $u(t,x)$ will be constructed. It will be done in several steps.

First of all we shall consider the case of the so called "factorial" trace. By this we mean that the boundary functions $\varphi_m(x)$ are such that the numbers $\varrho_m$ satisfy the inequalities

$$\varrho_m \leqq L^m \cdot m!, \quad m = 0,1,\ldots, \tag{2.5}$$

where $L > 0$ is a number.

Lemma 2.1. Let the inequalities (2.5) be fulfilled. Then there exists a function $u(t,x) \in W^{\infty}\{a_{n\alpha}, p_\alpha\}$ such that the conditions (2.2) are held.

Proof. The conditions (2.5) mean that the numbers $\varrho_m$ are the boundary values of an analytic function in a neighbourhood of zero, i. e. the series

$$z(t) = \sum_{m=0}^{\infty} \frac{\varrho_m}{m!} t^m$$

converges on some interval $[0,b]$. We may assume that $b \leqq a$.

Further, since the sequence of numbers $M_n^c$ satisfies the condition (2.4), then due to Lemma 1.1, Chapter I, for any $b > 0$ and $0 < q < 1$ there exists a function $\Phi(x) \in C_0^{\infty}(0,b)$ satisfying the following conditions:

1) $\Phi(0) = 1$, $D^k\Phi(0) = 0$, $k = 1,2,\ldots;$

2) $D^k\Phi(b) = 0$, $k = 0,1,\ldots;$

3) $|D^k\Phi(x)| \leqq K\, q^k\, M_k^c$, $k = 0,1,\ldots,$

where $K > 0$ is a constant, $x \in [0,b]$ is arbitrary. Moreover, the constant $K > 0$ depends on the choice of $b$ and $q$ but not on $k = 0,1,\ldots$

We now put

$$f(t,x) = \sum_{m=0}^{\infty} \frac{\varphi_m(x)}{m!} t^m, \quad x \in \mathbb{R}, \tag{2.6}$$

and define the function

$$u(t,x) = \begin{cases} f(t,x)\Phi(t), & t \in [0,b],\ x \in \mathbb{R}\ ; \\ 0 & , t \in [b,a],\ x \in \mathbb{R}\ , \end{cases}$$

where the function $\Phi(t)$ is defined by conditions 1) - 3) with $q < 1/2$. This function is the desired function.

First we note that, by virtue of (2.5) and the known inequalities of the Sobolev imbedding theorems (see S. L. SOBOLEV [1], S. M. NIKOLSKIJ [1], A. KUFNER, O. JOHN and S. FUČIK [1] et al.), the series (2.6) together with all its derivatives converges uniformly for $t \in [0,b]$ and $|x| \leq R$, where R is an arbitrary positive number. Consequently, the functions f(t,x) and u(t,x) are infinitely differentiable functions; moreover, the function u(t,x) satisfies the boundary conditions (2.2).

It remains to show that $\varrho(u) < \infty$, i. e. $u(t,x) \in W^{\infty}\{a_{n\alpha}, p\}(G)$. Using Leibniz's formula, we get that for any $t \in [0,b]$.

$$|D_t^n D_x^\alpha u(t,x)|_p \equiv \left| D_t^n \left[ D_x^\alpha f(t,x)\Phi(t)\right]\right|_p$$

$$\leq \sum_{k=0}^{n} \binom{k}{n} \left| D_t^k D_x^\alpha f(t,x) D_t^{n-k}\Phi(t)\right|_p .$$

From this inequality, taking into account the formula for f(t,x) and property 3) of $\Phi(t)$ (see Lemma 2.1), we find that

$$|D_t^n D_x^\alpha u|_p \leq K \sum_{k=0}^{n} \binom{k}{n} q^{n-k} M_{n-k}^c \sum_{m=0}^{\infty} |D_x^\alpha \varphi_m(x)|_p \times D_t^k \frac{t^m}{m!} .$$

Consequently,

$$\sum_{n+|\alpha|=0}^{\infty} a_{n\alpha}^{\frac{1}{p}} |D_t^n D_x^\alpha u|_p \leq K \sum_{n=0}^{\infty} \left\{ \sum_{k=0}^{n} \binom{k}{n} q^{n-k} M_{n-k}^c \right.$$

$$\left. \times \sum_{m=0}^{\infty} \left[ \sum_{|\alpha|=0}^{\infty} a_{n\alpha}^{\frac{1}{p}} |D_x^\alpha \varphi_m(x)|_p \cdot \max_{t \in (0,b)} D_t^k \frac{t^m}{m!} \right] \right\}, \tag{2.7}$$

where $K > 0$ is a constant.

From the inequality (2.7), taking into account the conditions of the theorem and the definition of numbers $\varrho_m$, we get that

$$\sum_{n+|\alpha|=0}^{\infty} a_{n\alpha}^{\frac{1}{p}} |D_t^n D_x^\alpha u|_p \leq K \sum_{n=0}^{\infty} \left[ \sum_{k=0}^{n} q^{n-k} M_{n-k}^c \right.$$

$$\left. \times \max_{t \in (0,b)} |D_t^k z(t)| \right] (M_n^c)^{-1} .$$

From this inequality, using the analyticity of z(t) and logarithmic convexity of the sequences n! and $M_n^c$, we find that

$$\sum_{n+|\alpha|=0}^{\infty} a_{n\alpha}^{\frac{1}{p}} \|D_t^n D_x^\alpha u\|_p \leq \sum_{n=0}^{\infty} \left[\sum_{k=0}^{n} q^{n-k} M_{n-k}^c L^k\, k!\right] (M_n^c)^{-1}$$

$$\leq K \sum_{n=0}^{\infty} \left[(2L)^n\, n! \sum_{k=0}^{n} 1_n^k\right] (M_n^c)^{-1}, \tag{2.8}$$

where $L > 0$ is a constant and $1_n = (M_n^c/n!)^{\frac{1}{n}} \cdot \frac{q}{L}$.

It is easy to see that from condition (2.4) it follows that $1_n \to 0$ as $n \to \infty$. Consequently, from the latter inequality we find that

$$\sum_{n+|\alpha|=0}^{\infty} a_{n\alpha}^{\frac{1}{p}} \|D_t^n D_x^\alpha u\|_p \leq K\left[(2L)^n\, n!\, 1_n^n\right] (M_n^c)^{-1} \equiv K \sum_{n=0}^{\infty} (2q)^n < \infty ,$$

since $q < 1/2$.

Thus, we have proved that $u(t,x) \in W^\infty\left\{a_{n\alpha}^{\frac{1}{p}}, 1\right\}(G)$ and, a fortiori, $u(t,x) \in W^\infty\{a_{n\alpha}, p\}(G)$. Q.E.D.

Let us return to the proof of the main theorem. First we note that without loss of generality one can assume that $r \leq a$. Further, it is evident that the desired function will be constructed if we construct a function $u(t,x) \in W^\infty\{a_{n\alpha}, p\}(G_b)$, where $G_b$ is the strip $\{0 \leq t \leq b,\ x \in \mathbb{R}^n\}$, satisfying the initial conditions (2.2) at $t = 0$ and the null conditions at $t = b$ (we recall that $b \leq a$).

For the construction of this function we indicate a suitable family of basic functions $v_m(t) \in C^\infty(0,b)$ satisfying the conditions $D_t^n v_m(0) = \delta_{mn}$, where $\delta_{mn}$ is the Kronecker delta. We then establish that the function

$$v(t,x) = \sum_{m=0}^{\infty} v_m(t)\varphi_m(x)$$

has a finite integral $\varrho(v)$, i. e. $v(t,x) \in W^\infty\{a_{n\alpha}, p\}(G)$.

It is obvious that this function satisfies the conditions (2.2) at $t = 0$. However, the function $v(t,x)$ does not satisfy the null conditions at $t = b$. After establishing that $v(t,x)$ has a "factorial" trace at $t = b$, we modify $v(t,x)$ with the use of Theorem 2.1 so that the resultant function satisfies both conditions (2.2).

1. Construction of basic functions. Let m be a non-negative integer and $b < r/2$ (see the condition 2) in Theorem 2.1). We choose the numerical sequence

$$b_{km} = \frac{M_m^c}{d_m^k M_{m+k}^c}, \quad k = 0,1,\ldots,$$

where $d_m = 4s_m/b$ and $s_m$ is defined by (2.3). It is easy to see that

$$\sum_{k=1}^{\infty} \frac{b_{km}}{b_{k+1,m}} = \frac{1}{d_m} \sum_{k=1}^{\infty} \frac{M^c_{m+k-1}}{M^c_{m+k}} < \frac{b}{3}.$$

Consequently, there exists a positive number $q < 1$ such that the numbers

$$\mu_{km} = \frac{1}{q} \frac{b_{km}}{b_{k-1,m}}, \quad k = 1,2,\ldots,$$

satisfy the conditions of Lemma 1.1, Chapter I. Therefore, there exists a family of functions $\Phi(t) \in C^\infty(0,b)$, satisfying all the conditions of this lemma. In particular, condition 3) leads to the estimates

$$\max_{t\in(0,b)} |D_t^k \Phi_m(t)| \leq \frac{q^k d_m^k M^c_{m+k}}{M^c_m}, \quad k = 0,1,\ldots \tag{2.9}$$

Now we put for $t \in (0,b)$

$$v_0(t) = \Phi(t), \quad v_1(t) = d_1 \int_0^{t/d_1} \Phi_1(\eta)\,d\eta,$$

$$v_m(t) = \frac{d_m}{(m-2)!} \int_0^t (t-\tau)^{m-2} \int_0^{\tau/d_m} \Phi_m(\eta)\,d\eta\,d\tau, \quad m = 2,3,\ldots$$

This sequence of functions is the desired basic family of functions. It is clear that the functions $v_m(t)$ satisfy the conditions $D_t^n v_m(0) = \delta_{mn}$, $n = 0,1,\ldots$ In addition, starting from the construction of $v_m(t)$ and (2.9), we get that for any $t \in (0,b)$

$$|D_t^n v_m(t)| \leq \frac{d_m b^{m-1-n}}{(m-1-n)!}, \quad \text{if } n \leq m-1; \tag{2.10}$$

$$|D_t^n v_m(t)| \leq q^{n-m} \frac{M^c_n}{M^c_m}, \quad \text{if } n \geq m. \tag{2.11}$$

2. Main extension. Now we want to construct a function $v(t,x)$, satisfying the basic conditions (2.2) at $t = 0$; viz. we put

$$v(t,x) = \sum_{m=0}^{\infty} v_m(t)\varphi_m(x)$$

and show that $v(t,x) \in W^\infty\{a_{n\alpha},p\}(G_b)$ (it is evident that $D_t^m v(0,x) = \varphi_m(x)$, $m = 0,1,\ldots$). Indeed, in virtue of the estimates (2.10), (2.11), for any $t \in (0,b)$ we get the inequality

$$|D_t^n D_x^\alpha v(t,x)|_p \leq \sum_{m=0}^{\infty} |D_t^n v_m(t)| \cdot |D_x^\alpha \varphi_m(x)|_p$$

$$\leq \sum_{m=0}^{n} q^{n-m} \frac{M_n^c}{M_m^c} |D_x^\alpha \varphi_m(x)|_p + \sum_{m=n+1}^{\infty} \frac{d_m b^{m-1-n}}{(m-1-n)!} |D_x^\alpha \varphi_m(x)|_p ,$$

where, we recall, $|\cdot|_p$ is the norm in the space $L_p$.

From this result, taking into account the definition of the numbers $\varrho_m$, we find that

$$\sum_{n+|\alpha|=0}^{\infty} a_{n\alpha}^{\frac{1}{p}} \|D_t^n D_x^\alpha v(t,x)\|_p$$

$$\leq K \sum_{n=0}^{\infty} \left[ \sum_{m=0}^{n} q^{n-m} (M_m^c)^{-1} \varrho_m + (M_n^c)^{-1} \sum_{m=n+1}^{\infty} \varrho_m \frac{s_m b^{m-1-n}}{(m-1-n)!} \right],$$

where $K > 0$ is a constant.

Further, in virtue of condition 2) of the theorem we have

$$1)\ \sum_{n=0}^{\infty} \sum_{m=0}^{n} q^{n-m} (M_m^c)^{-1} \varrho_m \leq \sum_{m=0}^{\infty} \varrho_m (M_m^c)^{-1} \sum_{k=0}^{\infty} q^k < \infty ;$$

$$2)\ \sum_{n=0}^{\infty} (M_n^c)^{-1} \sum_{m=n+1}^{\infty} \frac{s_m b^{m-1-n}}{(m-1-n)!} \leq \sum_{n=0}^{\infty} (M_n^c)^{-1} \frac{n!}{b^n}$$

$$\times \sum_{m=1}^{\infty} \varrho_m \frac{s_m (2b)^{m-1}}{(m-1)!} \leq K \sum_{n=0}^{\infty} (M_n^c)^{-1} \frac{n!}{b^n} < \infty ,$$

since the latter series also converges due to (2.4). Consequently, it follows from (2.13) that $v(t,x) \in W^\infty\{a_{n\alpha}^{1/p}, 1\}(G_b)$ and, a fortiori, $v(t,x) \in W^\infty\{a_{n\alpha}, p\}(G_b)$.

Remark. It is evident that the above arguments also give the locally uniferm convergence of the series (2.12) and all its derivatives.

3. Removal of the "residual" at $t = b$. The function constructed in step 2 does not satisfy the null conditions at $t = b$. Let us show, however, that its trace at $t = b$ is a factorial trace, i. e. this trace satisfies the inequalities (2.5).

Indeed, let $D_t^k v(b,x) \equiv \psi_k(t)$. Then

$$D_x^\alpha \psi_k(x) \equiv \sum_{m=k}^{\infty} D_t^k\, v_m(b) D_x^\alpha \varphi_m(x),$$

since $D_t^k\, v_m(b) = 0$ for $m \leq k-1$. From this, taking into account the estimates (2.10), we obtain that

$$|D_x^\alpha \psi_k(x)|_p \leq \sum_{m=k}^{\infty} |D_x^\alpha \varphi_m(x)|_p \cdot \frac{d_m b^{m-1-k}}{(m-1-k)!}.$$

Consequently, for any $n$

$$M_n^c \sum_{|\alpha|=0}^{\infty} a_{n\alpha}^{\frac{1}{p}} |D_x^\alpha \psi_k(x)|_p \leq \sum_{m=k}^{\infty} (M_n^c \sum_{|\alpha|=0}^{\infty} a_{n\alpha}^{\frac{1}{p}} |D_x^\alpha \varphi_m(x)|_p)$$

$$\times \frac{d_m b^{m-1-k}}{(m-1-k)!} \leq \sum_{m=k}^{\infty} \varrho_m \frac{d_m b^{m-1-k}}{(m-1-k)!}$$

$$\leq k!\, b^{-k} \sum_{m=k}^{\infty} \frac{d_m (2b)^{m-1}}{(m-1)!} \leq L^k\, k!,$$

where $L > 0$ is a constant (here again we use the condition 2) and the choice of $b < r/2$). It follows that the trace of $v(t,x)$ at $t = b$ is a factorial trace.

Using now Lemma 2.1, we choose a function $w(t,x) \in W^\infty\{a_{n\alpha}, p\}$ in the strip $G_b$ so that

$$D_t^m\, w(0,x) = 0,\ D_t^m\, w(b,x) = D_t^m\, v(b,x),\ m = 0,1,\ldots$$

Then, putting $u(t,x) \equiv v(t,x) - w(t,x)$, we obtain a function solving the problem for the strip $G_b$. It is evident, however, that the original problem for the strip $G = [0,a] \times \mathbb{R}^\nu$ is also solved, since it suffices to add to $u(t,x)$ an arbitrary function from the space $\mathring{W}^\infty\{a_{n\alpha}, p\}(G_{ba})$, where $G_{ba} = \{b \leq t \leq a,\ x \in \mathbb{R}^\nu\}$. Theorem 2.1 is proved.

Example. Let $\mathring{W}^\infty\{a_n, p\}(0,a)$ and $W^\infty\{a_\alpha, p\}(\mathbb{R}^\nu)$ be the nontrivial spaces on the interval $(0,a)$ and on the full Euclidean space (see § 1 and § 2, Chapter I). Then, as it is easy to see, the space $\mathring{W}^\infty\{a_{n\alpha}, p\}(G)$, where $a_{n\alpha} = a_n \cdot a_\alpha$, $G = [0,a] \times \mathbb{R}^\nu$ is also nontrivial. In this case it is clear that it is possible to put $b_n = a_n$. It is also clear that

$$\varrho_m = \sum_{|\alpha|=0}^{\infty} a_\alpha^{\frac{1}{p}} |D_x^\alpha \varphi_m(x)|_p.$$

Thus, the conditions of Theorem 2.1 take the form

1) $\varrho_m < \infty$;

2) $\sum_{m=0}^{\infty} \varrho_m \cdot \max\left(\frac{s_m r^m}{m!}, (M_m^c)^{-1}\right) < \infty$,

where $M_n^c$ is a convex regularization by means of logarithms of the sequence $a_n^{-\frac{1}{p}}$ and $r > 0$ is a positive number.

If the sequence $a_n^{-\frac{1}{p}}$ is a logarithmically convex sequence, then

$$M_m^c = a_m^{-\frac{1}{p}}, \quad s_m = \left(\frac{a_{m+1}}{a_m}\right)^{\frac{1}{p}} + \left(\frac{a_{m+2}}{a_{m+1}}\right)^{\frac{1}{p}} + \ldots$$

and condition 2) takes the form

$$\sum_{m=0}^{\infty} \varrho_m \cdot \max\left(\frac{s_m r^m}{m!}, a_m^{-\frac{1}{p}}\right) < \infty . \tag{2.14}$$

The latter inequality has a very simple form if $a_n$ is a rapidly decreasing sequence, viz. $a_{n+1} \leqq a_n^2$, $a_0 < 1$ $(n = 0,1,\ldots)$. In this case one can show (after simple calculations) that

$$s_m \leqq K\, a_m^{\frac{1}{p}}, \quad m = 0,1,\ldots,$$

where $K > 0$ is a constant. Consequently, for large m

$$\max\left(\frac{s_m r^m}{m!}, a_m^{\frac{1}{p}}\right) = a_m^{\frac{1}{p}}$$

and the inequality (2.14) takes the form

$$\sum_{m=0}^{\infty} \varrho_m a_m^{\frac{1}{p}} < \infty .$$

Remark. All considerations of this paragraph are also valid for the case of functions of one independent variable, i. e. in the case of interval $(0,a)$. In order to formulate the algebraic conditions for extension, it suffices (in the conditions of Theorem 2.1 and Lemma 2.1) to put $\varrho_m = |b_m|$, where $b_m = D_t^m u(0)$ are the boundary values of function $u(t)$ at $t = 0$.

In conclusion we note that in the case of rapidly decreasing coefficients $a_{n\alpha}$ some precise algebraic conditions of trace were obtained by G. S. BALASHOVA [1]. We here formulate the main result of her work for the case of one dimension.

Indeed, let

$$W^{\infty}\{a_n,p,r\} = \left\{u(x) \in C^{\infty}(0,a) : \sum_{n=0}^{\infty} a_n \| D^n u \|_r^p < \infty\right\}$$

be the Sobolev space of infinite order on the interval (0,a). We find a function $u(x) \in W^{\infty}\{a_n,p,r\}$ such that

$$D^n u(0) = b_n, \; D^n u(a) = c_n, \; n = 0,1,\ldots,$$

where $b_n$, $c_n$ are the given number sequences.

Theorem 2.2. Let the sequence $a_n$, $n = 0,1,\ldots$, be such that $a_{n+1} \leq a_n^2$, $a_0 < 1$. If for some number s

$$\sum_{n=1}^{\infty} a_n \left( |b_n|^{\frac{r-s}{r}p} + |b_{n-1}|^{\frac{r+s(r-1)}{r}p} \right) < \infty , \tag{2.15}$$

then there exists a function $u(x) \in W^{\infty}\{a_n,p,r\}$, satisfying the conditions

$$D^n u(0) = b_n, \; D^n u(a) = 0, \; n = 0,1,\ldots \tag{2.16}$$

The case $D^n u(a) = c_n$ is considered by analogy.

If $s = r$ the condition (2.15) takes the form

$$\sum_{n=1}^{\infty} a_n |b_{n-1}|^{rp} < \infty .$$

In particular, for $r = 1$ this condition is not only sufficient but is also necessary for the existence of some extension of the trace (2.16) in the space $W^{\infty}\{a_n,p,1\}$.

If $s = r/(r+1)$, the condition (2.15) takes the form

$$\sum_{n=1}^{\infty} a_n |b_n|^{\frac{r}{r+1}p} < \infty .$$

In particular, for $r = +\infty$ this condition is not only sufficient but is also necessary for the existence of some extension of the trace (2.16) in the space $W^{\infty}\{a_n,p,\infty\}$.

For $1 < r < \infty$ the condition (2.15) is in a certain sense final.

## § 3. Inhomogeneous Dirichlet problem of infinite order

In the region $G \subset \mathbb{R}^\nu$ with boundary $\Gamma$ the following inhomogeneous boundary value problem is considered:

$$\sum_{|\alpha|=0}^{\infty} (-1)^{|\alpha|} \; D^\alpha A_\alpha(x, D^\gamma u) = h(x), \quad |\gamma| \leq |\alpha| , \tag{3.1}$$

$$D^\omega u\big|_\Gamma = f_\omega(x'), \quad x' \in \Gamma , \quad |\omega| = 0,1,\ldots \tag{3.2}$$

Let us suppose that the conditions a) - c) from § 2, Chapter II, are fulfilled:

a) For any $x \in G$, $\xi_\gamma$ and $\eta_\alpha$, where $|\alpha| \leq m$, $|\gamma| \leq |\alpha|$ $(m = 0,1,\ldots)$, the following inequality is valid:

$$\left| \sum_{|\alpha|=0}^{m} A_\alpha(x,\xi_\gamma)\bar{\eta}_\alpha \right| \leq K \sum_{|\alpha|=0}^{m} a_\alpha |\xi_\alpha|^{p_\alpha - 1} \cdot |\eta_\alpha| + b,$$

where $K > 0$, $a_\alpha \geq 0$, $p_\alpha > 1$ and $b \geq 0$ are constants; moreover, the sequence $p_\alpha$ is bounded;

b) $$\operatorname{Re} \sum_{|\alpha|=0}^{m} A_\alpha(x,\xi_\gamma)\bar{\xi}_\alpha \geq \delta \sum_{|\alpha|=0}^{m} a_\alpha |\xi_\alpha|^{p_\alpha} - C$$

where $\delta > 0$ and $C > 0$ are constants;

c) The space $\mathring{W}^\infty\{a_\alpha, p_\alpha\}$ is nontrivial.

Moreover, we shall assume that the boundary values (3.2) admit the extension $f(x) \in W^\infty\{a_\alpha, p_\alpha\}$ into the domain G, i. e. there exists the function $f(x) \in W^\infty\{a_\alpha, p_\alpha\}$ such that

$$D^\omega f(x)\big|_\Gamma = f_\omega(x'), \quad x' \in \Gamma , \quad |\omega| = 0,1,\ldots$$

Definition 3.1. The function $u(x) \in W^\infty\{a_\alpha, p_\alpha\}$ is called a weak solution of problem (3.1), (3.2) if $u(x) - f(x) \in \mathring{W}^\infty\{a_\alpha, p_\alpha\}$ and the identity

$$\sum_{|\alpha|=0}^{\infty} (A_\alpha(x, D^\gamma u), D^\alpha v) = (h, v)$$

is valid for any function $v(x) \in \mathring{W}^\infty\{a_\alpha, p_\alpha\}$.

Theorem 3.1. Suppose conditions a) - c) are satisfied. Then, for any right side $h(x) \in W^{-\infty}\{a_\alpha, p'_\alpha\}$, $p'_\alpha = p_\alpha/(p_\alpha - 1)$ (see § 2, Chapter II) there exists at least one solution $u(x) \in W^\infty\{a_\alpha, p_\alpha\}$ of problem (3.1), (3.2).

Proof. The proof of this theorem is carried out in essentially the same way as in the case of homogeneous boundary conditions (Theorem 2.1, Chapter II). We therefore confine ourselves to a brief outline.

An approximate solution can be found in the form

$$u_m(x) = f(x) + z_m(x),$$

where $z_m(x)$ is a solution of the boundary value problem of order $2m+2$ $(m = 0,1,\ldots)$

$$\sum_{|\beta|=m+1} (-1)^{m+1} b_\beta D^{2\beta} z_m + \sum_{|\alpha|=0}^{m} (-1)^{|\alpha|} D^\alpha A_\alpha(x, D^\gamma(f + z_m)) = h_m(x), \tag{3.3}$$

$$D^\omega z_m\big|_\Gamma = 0, \quad |\omega| \leqq m. \tag{3.4}$$

Here $b_\beta \geqq 0$ is a sequence of numbers, tending to zero sufficiently rapidly as $m \to \infty$ and

$$h_m(x) = \sum_{|\alpha|=0}^{m} (-1)^{|\alpha|} a_\alpha D^\alpha h_\alpha(x)$$

is the partial sum of the series

$$h(x) = \sum_{|\alpha|=0}^{\infty} (-1)^{|\alpha|} a_\alpha D^\alpha h_\alpha(x).$$

It follows from the conditions of our theorem that the problem (3.3), (3.4) is solved for any $m = 0,1,\ldots$, i. e. there exists a family of solutions $z_m(x)$. Moreover, the following estimates

$$\sum_{|\beta|=m+1} b_\beta \| D^\beta z_m \|_2^2 + \sum_{|\alpha|=0}^{m} a_\alpha \| D^\alpha z_m \|_{p_\alpha}^{p_\alpha} \leqq K \tag{3.5}$$

are valid. The constant $K > 0$ in (3.5) depends on $f(x)$ and $h(x)$ but not on $z_m(x)$, $m = 0,1,\ldots$

From the latter inequalities there exists (as usual) a subsequence $z_k(x)$, $k = 0,1,\ldots$, of the sequence $z_m(x)$ which together with its derivatives converges uniformly to some function $z(x) \in \mathring{W}^\infty\{a_\alpha, p_\alpha\}$. The function $u(x) = f(x) + z(x)$ is a solution of the original problem (3.1), (3.2). The theorem is proved.

## CHAPTER IV

# SPACES $W^{\infty}\{a_\alpha,p\}$ AS LIMITS OF BANACH SPACES. EXAMPLES

## Introduction

In this chapter a general point of view on the spaces $W^{\infty}\{a_\alpha,p\}$ and $W^{-\infty}\{a_\alpha,p'\}$ as limits of Banach spaces is developed. For this aim in § 1 we introduce and investigate the limits

$$X_\infty = \lim_{m\to\infty} X_m,$$

where

$$X_1 \supset X_2 \supset \ldots \supset X_m \supset \ldots$$

and

$$X_{-\infty} = \lim_{m\to\infty} X_{-m},$$

where

$$X_{-1} \subset X_{-2} \subset \ldots \subset X_{-m} \subset \ldots$$

The main feature of our constructions is that they are closed constructions in the category of Banach spaces, since the limit spaces $X_\infty$ and $X_{-\infty}$ are also Banach spaces.

The dual chain

$$X_\infty \subset \ldots \subset X_m \subset \ldots \subset X_1 \subset X_{-1} \subset \ldots \subset X_{-m} \subset \ldots \subset X_{-\infty},$$

where $X_{-m}$ is the dual space of $X_m$, has, probably, the largest practical interest.

Roughly speaking, in the dual chain the space $X_{-\infty}$ is nontrivial if and only if the space $X_\infty$ is nontrivial. We give some conditions under which the formula

$$X_{-\infty} = (X_\infty)^*$$

is valid, i. e. we give such conditions under which the dual chain is the topologically closed chain. Some applications of the results obtained in § 1 are given in the second paragraph. In particular, we prove that for $p > 1$ the space $W^{-\infty}\{a_\alpha,p'\}$, $p' = p/(p-1)$, consists of the ultra-distributions $h(x)$ (and only such $h(x)$) of the following type

$$h(x) = \sum_{|\alpha|=0}^{\infty} a_\alpha D^\alpha h_\alpha(x),$$

where $h_\alpha(x) \in L_{p'}(G)$ and

$$\sum_{|\alpha|=0}^{\infty} a_\alpha \| h_\alpha(x) \|_{p'}^{p'} < \infty .$$

Thus, the space $W^{-\infty}\{a_\alpha, p'\}$ which was introduced in Chapter II as the space of right sides of a Dirichlet problem of infinite order is the topological dual space of the space $W^{\infty}\{a_\alpha, p\}$.

Some properties of $W^{\infty}\{a_\alpha, p\}$ and $W^{-\infty}\{a_\alpha, p'\}$ are established: separability, reflexivity, uniform convexity etc. Moreover, in these spaces there are classical Clarkson inequalities.

## § 1. Some limits of monotone sequences of Banach spaces. Examples

Let

$$X_1 \supset X_2 \supset \ldots \supset X_m \supset \ldots \tag{1.1}$$

be a decreasing sequence of Banach spaces with norms

$$\|x\|_1 \leq \|x\|_2 \leq \ldots \leq \|x\|_m \leq \ldots \tag{1.2}$$

Definition 1.1. The space

$$X_\infty \equiv \lim_{m\to\infty} X_m \overset{\text{def}}{=} \left\{ x \in \bigcap_{m=1}^{\infty} X_m : \|x\|_\infty \overset{\text{def}}{=} \lim_{m\to\infty} \|x\|_m < \infty \right\}$$

is called the limit of a decreasing sequence of Banach spaces (1.1)

It's quite possible that the limit space $X_\infty$ is trivial, i. e. it contains only zero. In the sequel we shall suppose that the space $X_\infty$ in nontrivial. We shall also suppose that the convergences in the spaces $X_m$ are in conformity, i. e. if

$$x_n \to x \text{ in } X_m$$

and

$$x_n \to x' \text{ in } X_{m+1},$$

then $x' = x$.

Theorem 1.1. If the spaces $X_m$ are Banach spaces, i. e. complete normed spaces, then $X_\infty$ is also Banach space.

Proof. Obviously, it is necessary to verify the completeness of space $X_\infty$. Indeed, let $x_n \in X_\infty$, $n = 1,2,\ldots$, be a fundamental sequence. This means that for any $\varepsilon > 0$ there exists a number N such that.

$$\|x_{n+p} - x_n\|_\infty < \varepsilon \text{ , } n > N, \; p > 0. \tag{1.3}$$

Consequently, for every $m = 1,2,\ldots$ a fortiori

$$\|x_{n+p} - x_n\|_m < \varepsilon \text{ , } n > N, \; p > 0, \tag{1.4}$$

i. e. the sequence $x_n$ is a fundamental sequence in the space $X_m$. Since the convergences in $X_m$ are in conformity (by hypothesis), there exists an element

$$x \in \bigcap_{m=1}^{\infty} X_m$$

so that $x_n \to x$ in every $X_m$, as $n \to \infty$.

Tending $p \to \infty$ we get from inequality (1.4) that

$$\|x - x_n\|_m < \varepsilon \text{ , } n > N; \; m = 1,2,\ldots$$

From this after $m \to \infty$ we have

$$\|x - x_n\|_\infty < \varepsilon \text{ , } n > N.$$

It follows that $x_n \to x$ in $X_\infty$, i. e. the space $X_\infty$ is complete. The theorem is proved.

Examples. 1) Let $L_{p_m} \equiv L_{p_m}(G)$ be a sequence of Lebesgue spaces. Without loss of generality we can assume that meas $G = 1$. Then for $p_1 \leqq p_2 \leqq \ldots \leqq p_m \leqq \ldots$ the family of inequalities

$$\|u\|_{p_1} \leqq \|u\|_{p_2} \leqq \ldots \leqq \|u\|_{p_m} \leqq \ldots$$

is valid.

It is easy to see that as $p_m \to \infty$

$$L_{p_1} \supset L_{p_2} \supset \ldots \supset L_{p_m} \supset \ldots \longrightarrow L_\infty,$$

where $L_\infty$ is the space of bounded measurable functions in G.

2) By analogy, for the classical Sobolev spaces we have

$$W^r_{p_1} \supset W^r_{p_2} \supset \dots \supset W^r_{p_m} \supset \dots \longrightarrow W^r_\infty,$$

where

$$W^r_\infty = \left\{ u(x): \ \| u \|_\infty \equiv \sum_{|\alpha| = 0}^{r} \ \underset{x \in G}{\operatorname{vrai\ max}} |D^\alpha u(x)| < \infty \right\}.$$

3) Let $u(x): G \to \mathbf{C}^1$ be a function, where $G \subset \mathbb{R}^n$ is a bounded domain with boundary $\Gamma$. Let us denote

$$\overset{\circ}{W}{}^m\{a_\alpha, p\} = \left\{ u(x): \ \|u\|^p_{m,p} \equiv \sum_{|\alpha|=0}^{m} a_\alpha \|D^\alpha u\|^p_p, \ D^\omega u|_\Gamma = 0, \ |\omega| < m \right\},$$

where $a_\alpha \geqq 0$, $p \geqq 1$. As $m \to \infty$ we have a chain of imbedded spaces

$$\overset{\circ}{W}{}^1\{a_\alpha, p\} \supset \overset{\circ}{W}{}^2\{a_\alpha, p\} \supset \dots \supset \overset{\circ}{W}{}^m\{a_\alpha, p\} \dots \longrightarrow \overset{\circ}{W}{}^\infty\{a_\alpha, p\},$$

where

$$\overset{\circ}{W}{}^\infty\{a_\alpha, p\} = \left\{ u(x) \in C_0^\infty(G): \ \|u\|^p_{\infty, p} \equiv \sum_{|\alpha|=0}^{\infty} a_\alpha \|D^\alpha u\|^p_p < \infty \right\}.$$

Thus, Sobolev spaces of infinite order may be considered as the limits of Sobolev spaces of finite order, i. e.

$$\overset{\circ}{W}{}^\infty\{a_\alpha, p\} = \lim_{m \to \infty} \overset{\circ}{W}{}^m\{a_\alpha, p\}.$$

Is is clear that one may say just the same about the spaces $W^\infty\{a_\alpha, p\}$ of functions when the domain of definition is the torus $T^n$, the whole Euclidean space $\mathbb{R}^n$ etc.

Let us now consider the dual construction of a limit of an increasing sequence of normed spaces.

Let

$$X_{-1} \subset X_{-2} \subset \dots \subset X_{-m} \subset \dots$$

be a sequence of normed spaces with norms

$$\|x\|_{-1} \geqq \|x\|_{-2} \geqq \dots \geqq \|x\|_{-m} \geqq \dots$$

We consider the following linear space

$$Y_{-\infty} = \left\{ x \in \bigcup_{m=1}^{\infty} X_{-m}; \ \|x\|_{-\infty} = \lim_{m \to \infty} \|x\|_{-m} \right\}.$$

The introduced "norm" $\|x\|_{-\infty}$ has, in general, a nontrivial kernel and, therefore, defines only a seminorm in the space $Y_{-\infty}$. In this case we consider the space $Y_{-\infty}$ as the space which is identically equal to its factor space. Thus, we get a normed space $Y_{-\infty}$ with norm $\|x\|_{-\infty}$. In particular, if $\|x\|_{-\infty} = 0$ for every element

$$x \in \bigcup_{m=1}^{\infty} X_{-m},$$

then $Y_{-\infty}$ is trivial. (We remark, however, that in the typical examples the kernel of $\|x\|_{-\infty}$ equals to zero. In particular, this holds for all the examples which will be given below.)

Definition 1.2. The $\|x\|_{-\infty}$-completion of the space $Y_{-\infty}$ is called the limit space of the increasing sequence (1.5) of Banach spaces. We denote this space as

$$X_{-\infty} = \lim_{m\to\infty} X_{-m}.$$

Examples. 4) Let, as in example 1), $L_{p_m}$ be Lebesgue spaces in the region $G \subset \mathbb{R}^n$, meas $G = 1$. We denote by $X_{-m} = L_{p'_m}$ $(p'_m = p_m/(p_m - 1))$ the dual space of $X_m = L_{p_m}$. It is obvious that

$$L_{p'_1} \subset L_{p'_2} \subset \ldots \subset L_{p'_m} \subset \ldots \longrightarrow L_1,$$

where $L_1$ is the classical Lebesgue space of summable functions. In other words,

$$\lim_{m\to\infty} L_{p'_m} = L_1 \text{ or, equivalently, } \lim_{p'_m\to 1} L_{p'_m} = L_1.$$

5) By analogy for Sobolev spaces

$$W^r_{p'_1} \subset W^r_{p'_2} \subset \ldots \subset W^r_{p'_m} \subset \ldots \longrightarrow W^r_1.$$

6) As it follows from example 3)

$$\mathring{W}^{\infty}\{a_\alpha, p\} = \lim_{m\to\infty} \mathring{W}^m\{a_\alpha, p\},$$

where

$$\mathring{W}^m\{a_\alpha, p\} = \Big\{u(x):\ \|u\|^p_{m,p} = \sum_{|\alpha|=0}^{m} a_\alpha \|D^\alpha u\|^p_p,\ D^\omega u\big|_\Gamma = 0,\ |\omega| \le m-1\Big\}.$$

Let us denote by $W^{-m}\{a_\alpha, p'\}$ the dual space of the space $\mathring{W}^m\{a_\alpha, p\}$, i. e. the space of bounded linear functionals

$$h:\ \mathring{W}^m\{a_\alpha, p\} \longrightarrow \mathbb{C}^1$$

with norm

$$\|h\|_{-m,p'} = \sup \frac{|\langle h,v\rangle|}{\|v\|_{m,p}}, \quad v \in \mathring{W}^m\{a_\alpha,p\}.$$

It is well known that

$$W^{-m}\{a_\alpha,p'\} = \left\{h(x):\ h(x) = \sum_{|\alpha|=0}^{\infty} D^\alpha h_\alpha(x)\right\},$$

where $h_\alpha(x) \in L_{p'}$.

Let us now consider the chain of dual spaces

$$W^{-1}\{a_\alpha,p'\} \subset W^{-2}\{a_\alpha,p'\} \subset \ldots \subset W^{-m}\{a_\alpha,p'\} \subset \ldots$$

It is easy to see that the normed space

$$Y_{-\infty} = \left\{h(x) \in \bigcup_{m=1}^{\infty} W^{-m}\{a_\alpha,p'\},\ \|h\|_{-\infty,p'} \equiv \lim_{m\to\infty} \|h\|_{-m,p'}\right\}$$

is nontrivial if the space $\mathring{W}^\infty\{a_\alpha,p\}$ is nontrivial. Indeed, any element

$$h(x) \in \bigcup_{m=1}^{\infty} W^{-m}\{a_\alpha,p'\}$$

is, obviously, a bounded linear functional on the space $\mathring{W}^\infty\{a_\alpha,p\}$; moreover, for every m the inequality

$$\|h\|_{-m,p'} \geq \|h\|_*$$

is valid (here $\|h\|_*$ is the norm in space $(\mathring{W}^\infty\{a_\alpha,p\})^*$).

Consequently,

$$\|h\|_{-\infty,p'} \geq \|h\|_*.$$

It follows that the norm $\|h\|_{-\infty,p'}$ has the trivial kernel, since the imbedding

$$\mathring{W}^\infty\{a_\alpha,p\} \subset \mathring{W}^m\{a_\alpha,p\}$$

is dense.

Thus, for every function $h(x) \in \bigcup_{m=1}^{\infty} W^{-m}\{a_\alpha,p'\}$, $h(x) \not\equiv 0$, the norm $\|h\|_{-\infty,p'} > 0$, i. e. the space $Y_{-\infty}$ is nontrivial.

Thereby, the space

$$W^{-\infty}\{a_\alpha,p'\} = \lim_{m\to\infty} W^{-m}\{a_\alpha,p'\}$$

is also nontrivial. (We should recall that the space $W^{-\infty}\{a_\alpha,p'\}$ is the completion of the space $Y_{-\infty}$; see Definition 1.2.)

Below in § 2 we shall establish that for $p > 1$ the space $W^{-\infty}\{a_\alpha, p'\}$ is the topological dual space of $\mathring{W}^\infty\{a_\alpha, p\}$; moreover, any element $h(x) \in W^{-\infty}\{a_\alpha, p\}$ has the form

$$h(x) = \sum_{|\alpha|=0}^{\infty} a_\alpha D^\alpha h_\alpha(x),$$

where $h_\alpha(x) \in L_{p'}(G)$ and

$$\sum_{|\alpha|=0}^{\infty} a_\alpha \| h_\alpha(x) \|_{p'}^{p'} < \infty .$$

The examples 1), 4) give us the dual chain

$$\ldots \subset L_{p_m} \subset L_{p_{m-1}} \subset \ldots \subset L_{p_1} \subset L_{p_1'} \subset \ldots \subset L_{p_{m-1}'} \subset L_{p_m'} \subset \ldots$$

(without loss of generality one can put $p_1 \geq 2$).

Likewise the dual chain ($p \geq 2$)

$$\ldots \subset \mathring{W}^m\{a_\alpha, p\} \subset \ldots \subset \mathring{W}^1\{a_\alpha, p\} \subset W^{-1}\{a_\alpha, p'\} \subset \ldots \subset W^{-m}\{a_\alpha, p'\} \subset \ldots$$

corresponds to examples 2), 5).

Let us now consider the general case. Namely, let

$$\ldots \subset X_m \subset \ldots \subset X_1 \subset X_{-1} \subset \ldots \subset X_{-m} \subset \ldots \qquad (1.6)$$

be a sequence of Banach spaces $X_m$ and its dual spaces $X_{-m} = (X_m)^*$. We shall assume that the dualities $\langle x', x \rangle_m$ in the pairs $(X_{-m}, X_m)$ are in conformity. The latter means that if $x' \in X_{-m_0}$, then for any $m \geq m_0$ and $x \in X_m$ the equality

$$\langle x', x \rangle_m = \langle x', x \rangle_{m_0}$$

is valid.

Let $X_\infty$ and $X_{-\infty}$ denote the corresponding limits, i. e.

$$X_\infty = \lim_{m\to\infty} X_m, \quad X_{-\infty} = \lim_{m\to\infty} X_{-m}.$$

Our aim is to learn in what case these spaces are nontrivial.

Theorem 1.2. Let the space $X_\infty$ be nontrivial. Then the space $X_{-\infty}$ is nontrivial too.

Proof. Let $x_0 \in X_\infty$ and $x_0 \neq 0$. We fix a number $m_0$ and consider $x_0$ as the element of $X_{m_0}$. Using a well known corollary of the Hahn-

Banach theorem, we find a bounded linear functional $x' \in X_{-m_0}$ such that $\langle x',x\rangle_{m_0} = 1$. Since the dualities in the chain (1.6) are in conformity, then $\langle x',x\rangle_m = 1$ for all $m \geq m_0$. It remains to remark that

$$\|x'\|_{-\infty} \overset{\text{def}}{=} \lim_{m\to\infty}\|x'\|_{-m} \geq \lim_{m\to\infty} \frac{\langle x',x_0\rangle_m}{\|x_0\|_m} = \frac{1}{|x_0|_\infty} > 0,$$

since $x_0$ is not equal to zero. The theorem is proved.

We obtain now a theorem of inverse type.

Theorem 1.3. Let one of the following conditions be fulfilled:

1) the spaces $X_m$, $m = 1,2,..,$ are reflexive;

2) the imbeddings $X_{m+1} \subset X_m$, $m = 1,2,...$ are compact.

Then the triviality of $X_{-\infty}$ follows from the triviality of the space $X$ .

Proof. Let $X_\infty = \{0\}$, i. e. the space $X_\infty$ is trivial. We shall show that for every $x' \in \bigcup_{m=1}^{\infty} X_{-m}$ the norm

$$|x'|_{-\infty} \overset{\text{def}}{=} \lim_{m\to\infty}\|x'\|_{-m} = 0.$$

It will follow that the space $X_{-\infty}$ is trivial.

Indeed, let $x' \in X_{-m_0}$, where $m_0$ is the least number among the numbers $m$ such that $m' \in X_{-m}$. Taking into account the conformity of the dualities, we have for all $m \geq m_0$

$$|x'|_{-m} \overset{\text{def}}{=} \sup_{x\in X_m} \frac{\langle x',x\rangle_m}{|x|_m} = \sup_{x\in X_m} \frac{\langle x',x\rangle_{m_0}}{|x|_m}.$$

Suppose that the norms do not tend to zero. Then we have the sequence $x_m \in X_m$, $m = 1,2,...,$ such that

$$\frac{\langle x',x_m\rangle_{m_0}}{|x_m|_m} \geq \varepsilon,$$

where $\varepsilon > 0$ is a positive number (strictly speaking, we have a subsequence of the sequence $x_m$, but it is immaterial). Equivalently

$$\langle x',y_m\rangle_{m_0} \geq \varepsilon, \quad |y_m|_m = 1, \tag{1.7}$$

where $y_m = x_m/\|x_m\|_m$.

Since $\|y_m\|_s \leq \|y_m\|_m$ for $s \leq m$, then in view of condition 1) or condition 2) we may consider the sequence $y_m$ (in general, only a subsequence) as tending to an element

$$y \in \bigcap_{m=1}^{\infty} X_m,$$

and converging in any $X_s$ ($s = 1,2,\ldots$) weakly. Moreover, it is obvious that from (1.7) we get

$$\langle x', y\rangle_{m_0} \geq \varepsilon, \quad \|y\|_\infty \leq 1.$$

Since the space $X_\infty$ is trivial, these relations are impossible. This contradiction proves our theorem.

Theorem 1.4. Let $X_\infty$ be a reflexive and strongly convex space together with its dual space $X_\infty^*$. Then the formula

$$X_{-\infty} = X_\infty^*$$

is valid.

Proof. Indeed, in view of the hypotheses of the theorem there exists a dual operator $J\colon X_\infty \to X_\infty^*$, acting homeomorphically (see J. L. LIONS [3]. JU. A. DUBINSKIJ and S. I. POHOSAEV [1] et al.). Since $\|\cdot\|_{-\infty} \geq \|\cdot\|_{X_\infty^*}$, then for any sequence $h_n \in \bigcup_{m=1}^{\infty} X_{-m}$, which is fundamental in the norm $\|\cdot\|_{-s}$, there exists one and only one element $x = \lim_{n\to\infty} J^{-1}(h_n)$ from $X$. Identifying every fundamental sequence $h_n \in \bigcup_{m=1}^{\infty} X_{-m}$ with the corresponding element $x \in X^*$, we get the assertion of the theorem. Theorem 1.4 is proved.

Thus, under the conditions of Theorem 1.4 the dual chain is naturally closed to a complete dual chain

$$X_\infty \subset \ldots \subset X_m \subset \ldots \subset X_1 \subset X_{-1} \subset \ldots \subset X_{-m} \subset \ldots \subset X_{-\infty}.$$

## § 2. Some properties of the spaces $W^\infty\{a_\alpha, p\}$ and $W^{-\infty}\{a_\alpha, p'\}$

In this paragraph, on the basis of the previous constructions, we study some properties of the spaces $W^\infty\{a_\alpha, p\}$ and $W^{-\infty}\{a_\alpha, p'\}$. To be definite we limit our considerations to the case of spaces $\mathring{W}^\infty\{a_\alpha, p\}$ and $W^{-\infty}\{a_\alpha, p'\}$ of functions defined in a bounded domain $G \subset \mathbb{R}^n$. Let us emphasize, however, that all the results, which will be obtained below, are also valid in the cases of torus $T^n$, the full Euclidean space $\mathbb{R}^n$ etc.

At first we shall study the structure of the space $W^{-\infty}\{a_\alpha, p'\}$.

Theorem 2.1. Let $p > 1$, $p' = p/(p-1)$ and

$$W^{-\infty}\{a_\alpha, p'\} = \lim_{m\to\infty} W^{-m}\{a_\alpha, p'\},$$

where $W^{-m}\{a_\alpha, p'\}$ is the dual space of $\overset{\circ}{W}{}^m\{a_\alpha, p\}$. Then there exists the following representation

$$W^{-\infty}\{a_\alpha, p'\} = \Big\{h(x) : h(x) = \sum_{|\alpha|=0}^{\infty} (-1)^{|\alpha|} a_\alpha D^\alpha h_\alpha(x),$$

where $h_\alpha(x) \in L_{p'}(G)$; moreover, $\displaystyle\sum_{|\alpha|=0}^{\infty} a_\alpha \|h_\alpha\|_{p'}^{p'} < \infty\Big\}$.

Proof. Let us recall that in view of Definition 1.2 the space $W^{-\infty}\{a_\alpha, p'\}$ is the completion of $\bigcup_{m=1}^{\infty} W^{-m}\{a_\alpha, p'\}$ in the norm

$$\|h\|_{-\infty,p'} \equiv \lim_{m\to\infty} \|h\|_{-m,p'},$$

where $\|h\|_{-m,p'}$ is the norm of $h(x)$ as a functional on the space $\overset{\circ}{W}{}^m\{a_\alpha, p\}$. Let us show that one and only one series

$$\sum_{|\alpha|=0} (-1)^{|\alpha|} a_\alpha D^\alpha h_\alpha(x),$$

noted in the theorem, corresponds to any element of this completion.

Indeed, let $h_s(x) \in \bigcup_{m=1}^{\infty} W^{-m}\{a_\alpha, p'\}$, $s = 1, 2, \ldots$, be a fundamental sequence in the norm $\|\cdot\|_{-\infty,p'}$. We note $m_0 = m_0(s)$ as the smallest number among all the numbers $m$ for which $h_s(x) \in W^{-m}\{a_\alpha, p'\}$. Then, as it is known from the classical theory of Sobolev spaces, the functions $h_s(x)$ can be represented in the form

$$h_s(x) = \sum_{|\alpha|=0}^{m} (-1)^{|\alpha|} a_\alpha D^\alpha h_{\alpha s}(x),$$

where $h_{\alpha s}(x) \in L_{p'}(G)$.

Let us now consider the family of Dirichlet problems of infinite order

$$L(u_s) = \sum_{|\alpha|=0}^{\infty} (-1)^{|\alpha|} D^\alpha(a_\alpha |D^\alpha u_s|^{p-2} D^\alpha u_s) = h_s(x), \qquad (2.1)_s$$

$$D^{\omega}u_s|_{\Gamma} = 0, \quad |\omega| = 0,1,\ldots \tag{2.2}_s$$

According to our results of § 1, Chapter II, the problem $(2.1)_s$, $(2.2)_s$ has the unique solution $u_s(x) \in \mathring{W}^{\infty}\{a_\alpha, p\}$; moreover for any $m \geqq m_0$ the formula

$$|u_s|^p_{\infty,p} \equiv \sum_{|\alpha|=0}^{\infty} a_\alpha |D^\alpha u_s|^p_p = \langle h_s, u_s\rangle_m$$

is valid.

Hence,

$$|u_s|^p_{\infty,p} \leqq \lim_{m\to\infty}(|h_s|_{-m,p'} \cdot |u_s|_{m,p}) = |h_s|_{-\infty,p'} \cdot |u_s|_{\infty,p},$$

or, equivalently

$$|u_s|_{\infty,p} \leqq |h_s|^{\frac{1}{p-1}}_{-\infty,p'} \leqq K < \infty , \tag{2.3}$$

since the sequence $h_s(x)$ is the fundamental sequence in the norm $|\cdot|_{-\infty,p'}$.

The estimate (2.3) implies that there exists a function $u(x) \in \mathring{W}^{\infty}\{a_\alpha, p\}$ such that $u_r(x) \to u(x)$ uniformly with all its derivatives.

Let us show that $u(x)$ is the unique limit point of the sequence $u_s(x)$. Indeed, let $u_k(x)$ be another subsequence of the sequence $u_s(x)$ so that

$$u_k(x) - w(x) \text{ in } C_0^{\infty}(G),$$

where $w(x)$ is a function from $\mathring{W}^{\infty}\{a_\alpha, p\}$. Let us prove that $w(x) \equiv u(x)$.

Obviously, we have

$$\langle L(u_r) - L(u_k), v\rangle = \langle h_r - h_k, v\rangle_m,$$

where $v(x) \in \mathring{W}^{\infty}\{a_\alpha, p\}$, $m \geqq \max[m_0(r), m_0(k)]$ (see the definition of number $m_0(r)$ at the beginning of our proof).

Hence,

$$\langle L(u_r) - L(u_k), v\rangle \leqq |h_r - h_k|_{-m,p'} \cdot |v|_{m,p};$$

from this, tending $m \to \infty$, we get the inequality

$$\langle L(u_r) - L(u_k), v\rangle \leqq |h_r - h_k|_{-\infty,p'} \cdot |v|_{\infty,p}.$$

Since the original sequence $h_k(x)$ is the fundamental sequence in the norm $|\cdot|_{-\infty,p'}$, then, tending now $k \to \infty$, $r \to \infty$, we immediately obtain that

$$\langle L(u) - L(w), v\rangle \leq 0,$$

where $v(x) \in \mathring{W}^{\infty}\{a_\alpha, p\}$ is an arbitrary function.

In view of uniqueness of the solution of the Dirichlet problem of infinite order (Theorem 1.1, § 1, Chapter II) $w(x) \equiv u(x)$. Q.E.D.

Thus, for every fundamental sequence $h_s(x)$, $s = 1,2,\ldots$, there exists one and only one function $u(x) \in \mathring{W}^{\infty}\{a_\alpha, p\}$ so that

$$u_s(x) = L^{-1}(h_s(x)) \longrightarrow u(x)$$

uniformly with all its derivatives. It is easy to see that if the fundamental sequences $h_s(x)$ and $h'_s(x)$ are equivalent, then the corresponding functions $u(x)$ and $u'(x)$ are equal, i. e. $u(x) \equiv u'(x)$. It remains to remark that for any $u(x) \in \mathring{W}^{\infty}\{a_\alpha, p\}$ the series $L(u)$ defines an element of the completion of $\bigcup_{m=1}^{\infty} W^{-m}\{a_\alpha, p'\}$ in the norm $|\cdot|_{-\infty,p'}$.

Therefore, the desired map

$$L(u) \longleftrightarrow \{h_s\}, \quad s = 1,2,\ldots$$

is defined. The theorem is proved.

Remark. In the case $p \geq 2$ one can give a simpler proof. It is sufficient in this case to use the inequality

$$|u - v|_{\infty,p} \leq K\|L(u) - L(v)\|_{-\infty,p'}^{\frac{1}{p-1}}, \quad K > 0,$$

which follows immediately from the inequality 2) of Proposition 1.1, Chapter II.

Corollary. There exists the equality

$$W^{-\infty}\{a_\alpha, p'\} = (\mathring{W}^{\infty}\{a_\alpha, p\})^*,$$

where $(\mathring{W}^{\infty}\{a_\alpha, p\})^*$ is the dual space of the space $\mathring{W}^{\infty}\{a_\alpha, p\}$.

Indeed, as we know, the operator $L(u)$ maps the space $\mathring{W}^{\infty}\{a_\alpha, p\}$ to the space $W^{-\infty}\{a_\alpha, p'\}$ one to one.

On the other hand, as it follows from the well known theory of monotone operators (see, for example, G. MINTY [1], F. E. BROWDER

[1], J. LERAY and J. L. LIONS [1] et al.) the same operator $L(u)$ defines a homeomorphism

$$L(u): \overset{\circ}{W}{}^{\infty}\{a_\alpha, p\} \longrightarrow (\overset{\circ}{W}{}^{\infty}\{a_\alpha, p\})^*,$$

since $L(u)$ is a strongly monotone operator. Therefore, the spaces mentioned in the corollary are the same. Q.E.D.

Theorem 2.2. The spaces $\overset{\circ}{W}{}^{\infty}\{a_\alpha, p\}$ and $W^{-\infty}\{a_\alpha, p'\}$ are separable.

Proof. At first, using the known fact of the separability of $W^{-m}\{a_\alpha, p'\}$, $m = 1, 2, \ldots$, we shall prove the separability of the space $W^{-\infty}\{a_\alpha, p'\}$. Namely, let $B_m$ be a countable set which is dense in $W^{-m}\{a_\alpha, p'\}$. It is evident that the set

$$B = \bigcup_{m=1}^{\infty} B_m$$

is a countable set too; moreover, the set $B$ is dense in $\bigcup_{m=1}^{\infty} W^{-m}\{a_\alpha, p'\}$ in the norm $\|\cdot\|_{-\infty, p'}$.

According to Theorem 2.1 any element $h(x) \in W^{-\infty}\{a_\alpha, p'\}$ has the form

$$h(x) = \sum_{|\alpha|=0}^{\infty} (-1)^{|\alpha|} a_\alpha D^\alpha h_\alpha(x) \equiv \lim_{m\to\infty} h_m(x),$$

where

$$h_m(x) = \sum_{|\alpha|=0}^{m} (-1)^{|\alpha|} a_\alpha D^\alpha h_\alpha(x) \in W^{-m}\{a_\alpha, p'\}.$$

Therefore, any element $h(x) \in W^{-\infty}\{a_\alpha, p'\}$ can be approximated by elements of the set $B$ with arbitrary small error. It follows that the space $W^{-\infty}\{a_\alpha, p'\}$ is a separable space too.

Let us now prove the separability of the space $\overset{\circ}{W}{}^{\infty}\{a_\alpha, p\}$. For this we remark that, as usual, it is easy to prove the density of the imbedding

$$\overset{\circ}{W}{}^{\infty}\{a_\alpha, p\} \subset \overset{\circ}{W}{}^{\infty}\{a_\alpha, q\}, \quad p \geq q,$$

(see Lemma 2.1, § 2, Chapter II). Therefore, it is enough to obtain the separability of $\overset{\circ}{W}{}^{\infty}\{a_\alpha, p\}$ for large $p$; namely, for $p \geq 2$. The latter is valid in view of homeomorphism

$$L(u): \overset{\circ}{W}{}^{\infty}\{a_\alpha, p\} \longleftrightarrow W^{-\infty}\{a_\alpha, p'\}, \quad p \geq 2,$$

and the separability of the space $W^{-\infty}\{a_\alpha, p'\}$ which was just proved. The theorem is proved.

It is known (see J. CLARKSON [1], S. L. SOBOLEV [1]) that in the space $L_p$ there are the following inequalities

$$\left|\frac{u+v}{2}\right|_p^q + \left|\frac{u-v}{2}\right|_p^q \leqq \frac{1}{2^{q-1}}\left(|u|_p^{\frac{q}{q-1}} + |v|_p^{\frac{q}{q-1}}\right)^{q-1}, \tag{2.4}$$

where $q = p/(p-1)$ if $1 < p \leqq 2$ and $q = p$ if $p \geqq 2$.

Let us obtain the analogons inequalities in the spaces $\mathring{W}^{\infty}\{a_\alpha, p\}$ and $W^{-\infty}\{a_\alpha, p'\}$.

<u>Theorem 2.3</u>. For any functions $u(x) \in \mathring{W}^{\infty}\{a_\alpha, p\}$ and $v(x) \in \mathring{W}^{\infty}\{a_\alpha, p\}$ there are the inequalities

$$\left|\frac{u+v}{2}\right|_{\infty,q}^q + \left|\frac{u-v}{2}\right|_{\infty,p}^q \leqq \frac{1}{2^{q-1}}\left(|u|_{\infty,p}^{\frac{q}{q-1}} + |v|_{\infty,p}^{\frac{q}{q-1}}\right)^{q-1}, \tag{2.5}$$

where $q = p/(p-1)$ if $1 < p \leqq 2$ and $q = p$ if $p \geqq 2$.

<u>Theorem 2.4</u>. For any functions $h(x) \in W^{-\infty}\{a_\alpha, p'\}$ and $f(x) \in W^{-\infty}\{a_\alpha, p'\}$ there are the inequalities

$$\left|\frac{h+f}{2}\right|_{-\infty,p'}^q + \left|\frac{h-f}{2}\right|_{-\infty,p'}^q \leqq \frac{1}{2^{q-1}}\left(|h|_{-\infty,p'}^{\frac{q}{q-1}} + \|f|_{-\infty,p'}^{\frac{q}{q-1}}\right)^{q-1}, \tag{2.6}$$

where $q = p'/(p'-1)$ if $1 < p' \leqq 2$ and $q = p'$ if $p' \geqq 2$.

<u>Proof of Theorem 2.3</u>. Let us consider only the case $1 < p \leqq 2$ (the case $p \geqq 2$ is considered by analogy). Taking into account the definition of the norm $|\cdot|_{\infty,p}$, we have

$$\left(\frac{1}{2}|u|_{\infty,p}^p + \frac{1}{2}|v|_{\infty,p}^p\right)^{\frac{1}{p-1}} \equiv \left[\frac{1}{2}\sum_{|\alpha|=0}^{\infty} a_\alpha(|D^\alpha u|_p^p + |D^\alpha v|_p^p)\right]^{\frac{1}{p-1}}$$

From this, applying the classical inequality (2.4) to the right side and using the inverse Minkowski inequality, we get

$$\left(\frac{1}{2}|u|_{\infty,p}^p + \frac{1}{2}|v|_{\infty,p}\right)^{\frac{1}{p-1}} \geqq \left[\sum_{|\alpha|=0}^{\infty} a_\alpha\left(\left\|\frac{D^\alpha u + D^\alpha v}{2}\right|_p^{\frac{p}{p-1}}\right.\right.$$

$$\left.\left. + \left|\frac{D^\alpha u - D^\alpha v}{2}\right|_p^{\frac{p}{p-1}}\right)^{p-1}\right]^{\frac{1}{p-1}} \geqq \left(\sum_{|\alpha|=0}^{\infty} a_\alpha\left|\frac{D^\alpha u + D^\alpha v}{2}\right|_p^p\right)^{\frac{1}{p-1}}$$

$$+ \left(\sum_{|\alpha|=0}^{\infty} a_\alpha\left|\frac{D^\alpha u - D^\alpha v}{2}\right|_p^p\right)^{\frac{1}{p-1}} \equiv$$

$$\equiv \left| \frac{D^\alpha u + D^\alpha v}{2} \right|_{\infty,p}^{\frac{p}{p-1}} + \left| \frac{D^\alpha u - D^\alpha v}{2} \right|_{\alpha,p}^{\frac{p}{p-1}}.$$

It is the desired inequality. The theorem is proved.

The proof of Theorem 2.4 is based on the unique solvability of Dirichlet problems of infinite order. Namely, in conformity with the main results of § 1, Chapter II, for any functions $h(x) \in W^{-\infty}\{a_\alpha, p'\}$ and $f(x) \in W^{-\infty}\{a_\alpha, p'\}$ there exist unique functions $u(x) \in \mathring{W}^\infty\{a_\alpha, p\}$ and $v(x) \in \mathring{W}^\infty\{a_\alpha, p\}$ $(p = p'/(p'-1))$ such that

$$h(x) = \sum_{|\alpha|=0}^{\infty} (-1)^{|\alpha|} D^\alpha (a_\alpha \varphi_\alpha(D^\alpha u)),$$

$$f(x) = \sum_{|\alpha|=0}^{\infty} (-1)^{|\alpha|} D^\alpha (a_\alpha \varphi_\alpha(D^\alpha v)),$$

where $\varphi(\xi) = |\xi|^{p-2}\xi$. In addition, it is easy to see

$$\|h\|_{-\infty,p'} = |u|_{\infty,p}^{p-1}, \ |f|_{-\infty,p'} = |v|_{\infty,p}^{p-1}. \tag{2.7}$$

Further, we have

$$|h \pm f|_{-\infty,p'} \equiv \sup_{|w|_{\infty,p} \leq 1} \langle h \pm f, w \rangle =$$

$$= \sup_{|w|_{\infty,p} \leq 1} \sum_{|\alpha|=0}^{\infty} \int_G \left[ a_\alpha \varphi(D^\alpha u) \pm a_\alpha \varphi(D^\alpha v) \right] D^\alpha w \, dx,$$

from which, using successively Hölder's inequality for the integrals and Hölder's inequality for the sums, we obtain the following inequality

$$|h \pm f|_{-\infty,p'} \leq \left( \sum_{|\alpha|=0}^{\infty} a_\alpha \int_G |\varphi(D^\alpha u) \pm \varphi(D^\alpha v)|^{p'} dx \right)^{\frac{1}{p'}}$$

$$\cdot \sup_{|w|_{\infty,p} \leq 1} \left( \sum_{|\alpha|=0}^{\infty} a_\alpha \int_G |D^\alpha w|^p dx \right)^{\frac{1}{p}}$$

$$\equiv \left( \sum_{|\alpha|=0}^{\infty} a_\alpha \int_G |\varphi(D^\alpha u) \pm \varphi(D^\alpha v)|^{p'} dx \right)^{\frac{1}{p'}} \tag{2.8}$$

From this inequality it follows that

$$\left| \frac{h+f}{2} \right|_{-\infty,p'}^{p'} + \left| \frac{h-f}{2} \right|_{-\infty,p'}^{p'} \leq$$

$$\leq \sum_{|\alpha|=0}^{\infty} a_\alpha\left(\left|\frac{\varphi(D^\alpha u)+\varphi(D^\alpha v)}{2}\right|_{p'}^{p'} + \left|\frac{\varphi(D^\alpha u)-\varphi(D^\alpha v)}{2}\right|_{p'}^{p'}\right).$$

Applying the classical Clarkson inequality (2.4) to the right side of the latter inequality, we find that

$$\left|\frac{h+f}{2}\right|_{-\infty,p'}^{p'} + \left|\frac{h-f}{2}\right|_{-\infty,p'}^{p'} \leq \frac{1}{2}\sum_{|\alpha|=0}^{\infty} a_\alpha\Big(|\varphi(D^\alpha u)|_{p'}^{p'}$$

$$+ |\varphi(D^\alpha v)|_{p'}^{p'}\Big) \equiv \frac{1}{2}\sum_{|\alpha|=0}^{\infty} a_\alpha\left(|D^\alpha u|_p^p + |D^\alpha v|_p^p\right)$$

$$\equiv \frac{1}{2}(|u|_{\infty,p}^{p} + |v|_{\infty,p}^{p}).$$

Taking into account formula (2.7) we get now

$$\left|\frac{h+f}{2}\right|_{-\infty,p'}^{p'} + \left|\frac{h-f}{2}\right|_{-\infty,p'}^{p'} \leq \frac{1}{2}|h|_{-\infty,p'}^{p'} + \frac{1}{2}|f|_{-\infty,p'}^{p'}.$$

This is the desired inequality (2.6) for the case $p' \geq 2$. The case $1 \leq p' \leq 2$ is considered by analogy. The theorem is proved.

Corollary. It is known the Clarkson inequalities imply the uniform convexity of the corresponding Banach spaces. Thus, the spaces $\mathring{W}^\infty\{a_\alpha, p\}$ and $W^{-\infty}\{a_\alpha, p'\}$ for $p > 1$, $p' > 1$ are uniform convex spaces. In addition, as it is known, the dual property of uniform convexity is the property of uniform smoothness of a Banach space (see, for example, N. BOURBAKI [1], M. M. DAY [1] et al.). Thereby, the spaces $\mathring{W}^\infty\{a_\alpha, p\}$ and $W^{-\infty}\{a_\alpha, p'\}$ are uniformly smooth spaces. Since any of the noted properties implies the reflexivity of the space, we conclude that the spaces $\mathring{W}^\infty\{a_\alpha, p\}$ and $W^{-\infty}\{a_\alpha, p'\}$ are reflexive if $p > 1$, $p' > 1$.

CHAPTER V

# IMBEDDING THEORY

## Introduction

In this Chapter some results of the imbedding theory of Sobolev spaces of infinite order are considered. It should be mentioned that the abstract questions of imbedding theory do not depend on the concrete structure of the spaces $W^{\infty}\{a_\alpha, p\}$ (as, on the contrary, takes place, for example, in Chapters I, III). Moreover, the criteria for imbedding (and for compact imbedding) of the spaces $W^{\infty}\{a_\alpha, p\}$ ($p_\alpha \equiv p$) are the special cases of the criteria of imbedding of the abstract limit spaces $X_\infty$ which were introduced in the previous Chapter.

In § 1 necessary and sufficient conditions for imbedding and compactly imbedding $X_\infty \subset Y_\infty$ are obtained. It is necessary to remark that the conditions of these criteria are given in terms of the asymptotic behaviour of the norms of the imbedding operators

$$i_{r,m}: X_r \longrightarrow Y_m$$

(see Theorem 1.2 in § 1). Thereby, the study of the concrete imbeddings

$$W^{\infty}\{a_\alpha, p\} \subset W^{\infty}\{b_\alpha, q\}$$

rests on the difficult and unknown (at this moment) problem of the exact estimates of the norms of the imbeddings of classical Sobolev spaces of finite order, as their orders tend to infinity. Thus, together with the general functional criteria, the imbedding conditions, which are expressed in "algebraic" terms, become very important. In particular, it is useful to have the conditions in terms of direct parameters $a_\alpha$, $p$, $b_\alpha$, $q$ of the spaces $W^{\infty}\{a_\alpha, p\}$ and $W^{\infty}\{b_\alpha, q\}$, of the characteristic functions of these spaces etc. These conditions are only sufficient, but they can be verified easily. These questions are investigated in § 3. Namely, we obtain simple criterion for the imbedding

$$W^{\infty}\{a_\alpha, 2\} \subset W^{\infty}\{b_\alpha, 2\}$$

(the case $p \neq 2$ is open).

The problem of the algebraic imbedding conditions of Sobolev spaces of infinite order is studied much better in the one-dimensional case, that is, in the case of spaces $W^{\infty}\{a_n,p\}$ of functions of one real variable. These results are given in § 4.

## § 1. Imbedding criterion for limits of Banach spaces

Let

$$X_1 \supset X_2 \supset \ldots \supset X_r \supset \ldots$$

be a sequence of imbedding Banach spaces and let $X_\infty = \lim_{r\to\infty} X_r$ be the limit of this sequence. We recall that

$$X_\infty = \left\{ u \in \bigcap_{r=1}^{\infty} X_r : \ |u|_\infty \overset{\text{def}}{=} \lim_{r\to\infty} |u|_r < \infty \right\},$$

where $|u|_1 \leqq |u|_2 \leqq \ldots$ are the norms in $X_1, X_2, \ldots$ (see Chapter IV).

Likewise, we consider the Banach space $Y_\infty = \lim_{m\to\infty} Y_m$, where

$$Y_1 \supset Y_2 \supset \ldots \supset Y_m \supset \ldots$$

is a second sequence of decreasing Banach spaces.

We are interested in the question of the existence of the imbedding operator

$$i_{\infty,\infty} : X_\infty \longrightarrow Y_\infty.$$

It is clear that the problem of imbedding $X_\infty \subset Y_\infty$ is deep only if both of these spaces are nontrivial. Thus, we shall further assume that the spaces $X_\infty$ and $Y_\infty$ are nontrivial.

Theorem 1.1 (unconditional criterion). The space $X_\infty$ can be imbedded in $Y_\infty$ if and only if the following conditions are satisfied:

1) for every $m \geqq 0$ the imbeddings

$$i_{\infty,m} : X \longrightarrow Y_m$$

are valid;

2) the limit

$$\lim_{m\to\infty} \|i_{\infty,m}\| = M, \tag{1.1}$$

where $\|i_{\infty,m}\|$ is the norm of the operator $i_{\infty,m}$, exists and is finite. Moreover,

$$\|i_{\infty,\infty}\| = \lim_{m\to\infty} \|i_{\infty,m}\| . \tag{1.2}$$

Proof. The sufficiency of conditions 1), 2) is evident. Indeed, if $u \in X_\infty$, then in view of condition 1) $u \in Y_m$ for all $m = 1,2,\ldots$ too; moreover,

$$\|u\|_{Y_m} \leq \|i_{\infty,m}\| \cdot \|u\|_{X_\infty}.$$

Tending $m \to \infty$ and taking into account the condition 2) we get the inequality

$$\|u\|_Y \leq \lim_{m\to\infty} \|i_{\infty,m}\| \cdot \|u\|_{X_\infty}.$$

It follows that the imbedding $X_\infty \subset Y_\infty$ is valid.

Necessity. Let the imbedding $X_\infty \subset Y_\infty$ be fulfilled. Then, a fortiori, there exists the imbedding $X_\infty \subset Y_m$, where $m = 1,2,\ldots$ is arbitrary. Therefore, it remains to establish the condition 2).

Let us assume the contrary, i. e.

$$\lim_{m\to\infty} \|i_{\infty,m}\| = +\infty. \tag{1.3}$$

The contradiction will be, obviously, obtained if we find an element $u \in X_\infty$ such that $\lim_{m\to\infty} \|u\|_{Y_m} = \infty$.

For this let us choose a family of numbers $N_k \to \infty$, $k = 0,1,\ldots,$ and consider the open unit ball

$$S(0,1) = \left\{ u \in X_\infty : \|u\|_{X_\infty} < 1 \right\}$$

in the space $X_\infty$. The condition (1.3) means that there exists a number $m_0$ and an element $u_0 \in S(0,1)$ such that $\|u_0\|_{Y_{m_0}} > N_0$.

In view of the continuity of the imbedding operator $i_{\infty,m_0}$ there exists a ball $S(u_0,\varepsilon_0) \subset S(0,1)$ with the center $u_0$ of a radius $\varepsilon_0 > 0$ such that the inequality

$$\|u\|_{Y_{m_0}} > \frac{N_0}{2}, \quad u \in S(u_0,\varepsilon_0),$$

is valid.

Since the sequence of bounded linear operators is simultaneously bounded or unbounded on every ball, then the condition (1.3) implies that there exists a number $m_1$ and an element $u_1 \in S(u_1,\varepsilon_0)$ such that $\|u_1\|_{Y_{m_1}} > N_1$. Now, taking into account the continuity of the operator $i_{\infty,m_1}$, we find a ball $S(u_1\,\varepsilon_1) \subset S(u_0,\varepsilon_0)$ (the notation is clear) such that the inequality

$$|u|_{Y_{m_1}} > \frac{N_1}{2}, \quad u \in S(u_1, \varepsilon_1),$$

is valid.

In the same way we find a family of imbedded balls

$$S(u_0, \varepsilon_0) \supset S(u_1, \varepsilon_1) \supset \ldots \supset S(u_k, \varepsilon_k) \supset \ldots$$

such that for any $u \in S(u_k, \varepsilon_k)$ the inequalities

$$|u|_{Y_{m_s}} > \frac{N_s}{2}, \quad s = 0,1,\ldots,k, \tag{1.4}$$

where $m_s \to \infty$ as $s \to \infty$, are valid. In particular, for the centers of the balls $u_k \in S(u_k, \varepsilon_k)$ these inequalities

$$|u_k|_{Y_{m_s}} > \frac{N_s}{2}, \quad s = 0,1,\ldots,k,$$

are valid too.

Without loss of generality one can assume that the family of balls $S(u_k, \varepsilon_k)$ tends to an element $u \in X_\infty$.

On the other hand, if $k \to \infty$, it follows from (1.4) that for any s

$$|u|_{Y_{m_s}} > \frac{N_s}{2},$$

where $N_s \to \infty$, as $s \to \infty$. As it was already remarked, this proves the necessity of our conditions 1) and 2). The validity of formula (1.2) is evident. The theorem is proved.

The criterion, proved in Theorem 1.1, is defined by the behaviour of the norms of imbedding operators

$$i_{\infty,m}: X_\infty \to Y_m, \quad m = 1,2,\ldots,$$

i. e. in this criterion the limit space is used.

It is also natural to have a criterion for the imbedding $X_\infty \subset Y_\infty$ in which only the spaces $X_r$ and $Y_m$ are used. For this we suppose that the spaces $X_r$ are reflexive and, in addition, the following condition is fulfilled:

For any $m > 0$ there exists a number $r(m) > 0$ such that the imbeddings

$$i_{r,m}: X_r \to \dot{Y}_m \tag{1.5}$$

are valid; moreover, these imbeddings are compact.

Under this condition the following Theorem holds:

Theorem 1.2 (conditional criterion). The space $X_\infty$ can be imbedded in $Y_\infty$ if and only if the limit $\lim\limits_{m\to\infty}\lim\limits_{r\to\infty}|i_{r,m}|$ exists and is finite. Moreover,

$$|i_{\infty,\infty}| = \lim_{m\to\infty}\lim_{r\to\infty}|i_{r,m}|. \qquad (1.6)$$

Proof. The sufficiency of the condition of our theorem is evident. As for its necessity, the proof of this fact is reduced to the proof of the necessity of the condition 2) of Theorem 1.1. Indeed, there exists the following

Lemma 1.1. Let the condition (1.5) be fulfilled. Then for any m there exists the formula

$$\lim_{r\to\infty}|i_{r,m}| = |i_{\infty,m}|,$$

where

$$i_{\infty,m}: X_\infty \longrightarrow Y_m.$$

Proof. Let $u \in X_\infty$. Then, a fortiori, $u \in X_r$, $r = 1,2,\ldots$ and, therefore, from (1.5) we have

$$|u|_{Y_m} \leqq |i_{r,m}| \cdot |u|_{X_r}, \quad r \geqq r(m).$$

Tending $r \to \infty$ and taking into account that the norms $|i_{r,m}|$ do not increase, we get the inequality

$$|u|_{Y_m} \leqq \lim_{r\to\infty}|i_{r,m}|\cdot|u|_{X_\infty}.$$

From this, it follows immediately that

$$|i_{\infty,m}| \leqq \lim_{r\to\infty}|i_{r,m}|. \qquad (1.7)$$

Let us show that, in fact

$$|i_{\infty,m}| = \lim_{r\to\infty}|i_{r,m}|.$$

Indeed, if we assume that

$$|i_{\infty,m}| < \lim_{r\to\infty}|i_{r,m}|,$$

then for all sufficiently large r there are some elements $u_r \in X_r$ such that

$$|u_r|_{X_r} \leqq 1, \quad |u_r|_{Y_m} \geqq |i_{\infty,m}| + \varepsilon, \tag{1.8}$$

where $\varepsilon > 0$ is a fixed number.

Since the spaces $X_r$ are reflexive, one can suppose (choosing a subsequence if needed) that in every $X_s$, $s = 1,2,\ldots$, the sequence $u_r$ $(r \geqq s)$ converges to an element $u \in \bigcap_{s=1}^{\infty} X_s$ weakly.

Further, since in view of (1.8) $|u_r|_{X_s} \leqq 1$ for $r \geqq s$, then $|u|_{X_s} \leqq 1$ for every $s = 1,2,\ldots$, too. It follows that $u \in X_\infty$ and $|u|_{X_\infty} \leqq 1$.

On the other hand, taking into account the compactness of operator

$$i_{r,m}: X_r \longrightarrow Y_m$$

we get that there exists a subsequence (we denote it as $u_r$ too) and an element $u$ such that $u_r$ converges to $u$ in $Y_m$ strongly, i. e. in the norm of $Y_m$.

Consequently, passing to the limit in the second inequality (1.8), as $r \to \infty$, we obtain that

$$|u|_{Y_m} \geqq |i_{\infty,m}| + \varepsilon,$$

where $\varepsilon > 0$ is a number. The latter inequality is impossible in view of the definition of the norm of the operator $i_{\infty,m}$. Lemma 1.1 and, therefore, Theorem 1.2 are proved.

Remark. It is evident, the property of the reflexivity of $X_r$ and the compactness of the imbeddings (1.5) were used only for the proof of necessity. Sufficiency is true without these conditions, i. e. the imbedding $X_\infty \subset Y_\infty$ is true if there exist only the imbeddings (1.5) and the limit, mentioned in Lemma 1.1, is finite.

Let us turn to the question of compactness of the imbedding operator

$$i_{\infty,\infty}: X_\infty \longrightarrow Y_\infty.$$

We recall that the operator $i_{\infty,\infty}$ is compact if it maps the unit ball of the space $X_\infty$ into a compact set of the space $Y_\infty$, i. e. if the unit ball in $X_\infty$ is a compact set in the norm of $Y_\infty$.

Let us suppose that the following condition is fulfilled:

For any $m > 0$ the imbedding $X_\infty \subset Y_m$ is compact. Under this condition there is the following

Theorem 1.3. The imbedding

$$i_{\infty,\infty}: X_\infty \longrightarrow Y_\infty$$

is compact if and only if the convergence

$$|u|_{Y_\infty} = \lim_{m\to\infty} |u|_{Y_m} \tag{1.9}$$

is uniform on the unit ball of the space $X_\infty$.

Proof. For the proof of the sufficiency we remark that in conformity with our assumption the unit ball $S \subset X_\infty$ is compact in $Y_m$, where $m = 1,2,\ldots,$ is arbitrary. Therefore, there exists a sequence $u_n \in S$, $n = 1,2,\ldots,$ and an element $u \in \bigcap_{m=1}^{\infty} Y_m$ such that $u_n \to u$ in $Y_m$ for all m. Since for any m

$$|u_n|_{Y_m} \leqq |u_n|_{Y_\infty} \leqq \| i_{\infty,\infty} \| \cdot \| u_n \|_{X_\infty} \leqq \| i_{\infty,\infty} \|, \tag{1.10}$$

then $u \in Y_\infty$. It remains to show that $u_n \to u$ in $Y_\infty$. Indeed, in view of (1.9) and the inequality (1.10), for any $\varepsilon > 0$ there exists a number $m > 0$ so that for any $n = 1,2,\ldots$

$$|u - u_n|_{Y_\infty} \leqq |u - u_n|_{Y_m} + \varepsilon .$$

From this for large n

$$|u - u_n|_Y \leqq 2\varepsilon ,$$

since, as it was remarked, $u_n \to u$ in $Y_m$ for any fixed m. The latter inequality means that $u_n \to u$ in $Y_\infty$. The sufficiency is proved.

The necessity will be proved by contradiction. Namely, let us suppose that the imbedding $i_{\infty,\infty}$ is compact but the condition (1.9) is false. It means that there exist a number $\varepsilon_0 > 0$, a subsequence of natural numbers (without loss of generality one can assume that this subsequence is the whole series of natural numbers) and a sequence of elements $u_m \in S$ such that

$$\|u_m\|_{Y_\infty} - \|u_m\|_{Y_m} \geqq \varepsilon_0 . \tag{1.11}$$

On the other hand, since the imbedding $i_{\infty,\infty}$ is compact, the sequence $u_m \in S$ may be considered as a convergent sequence in $Y_\infty$. Let $u \in Y_\infty$ be the limit element, i. e.

$$|u - u_m|_{Y_\infty} \to 0, \quad m \to \infty .$$

Consequently, if $m \to \infty$, then

$$|u_m|_{Y_\infty} - |u_m|_{Y_m} \leqq |u|_{Y_\infty} - |u|_{Y_m} + |u - u_m|_{Y_\infty} + \|u - u_m\|_{Y_m}$$

$$\leqq |u|_{Y_\infty} - |u|_{Y_m} + 2\|u - u_m|_{Y_\infty} \to 0.$$

The latter is, however, impossible in view of (1.11). The necessity and, therefore, Theorem 1.3 are proved.

## § 2. Applications to the spaces $W^\infty\{a_\alpha,p\}$

In this paragraph we apply the previous results to Sobolev spaces $W^\infty\{a_\alpha,p\}$ of infinite order. The abstract theory which was developed in § 1 may be applied to any type of the spaces $W^\infty\{a_\alpha,p\}$: $W^\infty\{a_\alpha,p\}$ (or $\mathring{W}^\infty\{a_\alpha,p\}$) of functions $u(x)\colon G \to \mathbb{C}^1$, where $G \subset \mathbb{R}^n$; $W^\infty\{a_\alpha,p\}$ $(T^n)$ of functions $u(x)\colon T^n \to \mathbb{C}^1$, where $T^n$ is torus; $W^\infty\{a_\alpha,p\}$ $(\mathbb{R}^n)$ of functions $u(x)\colon \mathbb{R}^n \to \mathbb{C}^1$ etc.

To be quite definite, we shall consider only the case of $W^\infty\{a_\alpha,p\}$, the domain of definition of which is a bounded domain in $\mathbb{R}^n$. Namely, let

$$W^\infty\{a_\alpha,p\} = \left\{u(x) \in C^\infty(G)\colon \ |u|^p_{a,} = \sum_{|\alpha|=0}^{\infty} a_\alpha \|D^\alpha u\|^p_p < \infty\right\},$$

$$W^\infty\{b_\alpha,q\} = \left\{u(x) \in C^\infty(G)\colon \ |u|^q_{b,\infty} = \sum_{|\alpha|=0}^{\infty} b_\alpha \|D^\alpha u\|^q_q < \infty\right\}$$

be the pair of Sobolev spaces of infinite order. Here $a_\alpha \geqq 0$, $b_\alpha \geqq 0$ are some nonnegative sequences; $p \geqq 1$, $q \geqq 1$ are some numbers. We shall assume that $a_o > 0$, $b_o > 0$.

<u>Remark.</u> It is evident that there is no question of nontriviality for the spaces introduced now. However, this question arises immediately in the cases of $\mathring{W}^\infty\{a_\alpha,p\}$, $W^\infty\{a_\alpha,p\}$ $(T^n)$, $W^\infty\{a_\alpha,p\}$ $(\mathbb{R}^n)$ etc. (see Chapter I). Of course, in these cases we assume that the corresponding spaces will be nontrivial.

We are interested in the question of the imbedding of the space $W^\infty\{a_\alpha,p\}$ in the space $W^\infty\{b_\alpha,q\}$. For the study of this question we apply the results of § 1 of the present Chapter. In view of it we recall that

$$W^\infty\{a_\alpha,p\} = \lim_{r\to\infty} W^r\{a_\alpha,p\},$$

where

$$W^r\{a_\alpha, p\} = \left\{u(x):\ |u|^p_{a,r} = \sum_{|\alpha|=0}^{r} a_\alpha |D^\alpha u|^p_p < \infty\right\}.$$

By analogy,

$$W^\infty\{b_\alpha, q\} = \lim_{m\to\infty} W^m\{b_\alpha, q\},$$

where

$$W^m\{b_\alpha, q\} = \left\{u(x):\ |u|^q_{b,m} = \sum_{|\alpha|=0}^{m} b_\alpha |D^\alpha u|^q_q < \infty\right\}.$$

Let

$$i_{\infty,\infty}:\ W^\infty\{a_\alpha, p\} \to W^\infty\{b_\alpha, q\}$$

be the imbedding operator. Then from Theorem 1.1 we get the following result.

Theorem 2.1 (unconditional criterion). The space $W^\infty\{a_\alpha, p\}$ can be imbedded in the space $W^\infty\{b_\alpha, q\}$ if and only if the following conditions are satisfied:

1) for any $m > 0$ there exists the imbedding

$$i_{\infty,m}:\ W^\infty\{a_\alpha, p\} \to W^m\{b_\alpha, q\};$$

2) the limit $\lim_{m\to\infty} |i_{\infty,m}|$ exists and is finite.

In this connection,

$$\|i_{\infty,\infty}\| = \lim_{m\to\infty} \|i_{\infty,m}\|.$$

Let us turn to the conditional criterion for the imbedding of our spaces. Namely, suppose that the spaces $W^r\{a_\alpha, p\}$ are reflexive and for any $m > 0$ there exists a number $r(m)$ such that for all $r > r(m)$ the imbeddings

$$i_{r,m}:\ W^r\{a_\alpha, p\} \to W^m\{b_\alpha, q\} \tag{2.1}$$

are valid; moreover, these imbeddings are compact. Under these conditions Theorem 1.2 implies the following

Theorem 2.2 (conditional criterion). Let condition (2.1) be satisfied. Then the space $W^\infty\{a_\alpha, p\}$ can be imbedded in the space $W^\infty\{b_\alpha, q\}$ if and only if there exists the finite limit

$$\lim_{m\to\infty} \lim_{r\to\infty} |i_{r,m}|.$$

In this connection

$$|i_{\infty,\infty}| = \lim_{m\to\infty} \lim_{r\to\infty} |i_{r,m}| .$$

Let us consider now the conditions under which the imbedding operator $i_{\infty,\infty}$ is compact. In conformity with the text of § 1 we assume that for any m the imbedding

$$W^{\infty}\{a_\alpha, p\} \subset W^{m}\{b_\alpha, q\} \tag{2.2}$$

is compact. Then there exists

Theorem 2.3. Let condition (2.2) be satisfied. Then the imbedding

$$W^{\infty}\{a_\alpha, p\} \subset W^{\infty}\{b_\alpha, q\}$$

is compact precisely when

$$\lim_{m\to\infty} \sum_{|\alpha|=m}^{\infty} b_\alpha \|D^\alpha u\|_q^q = 0$$

uniformly on the unit ball of the space $W^{\infty}\{a_\alpha, p\}$.

The theorem is a direct corollary of Theorem 1.3.

Remark. As it is known (see S. L. SOBOLEV [1], S. M. NIKOLSKIJ [1] et al.) in the case of a bounded domain the imbeddings (2.2) are always valid when the spaces $W^{r}\{a_\alpha, p\}$ and $W^{m}\{b_\alpha, q\}$ coincide with the classical Sobolev spaces $W_p^r$ and $W_q^m$. However, in the case of arbitrary coefficients $a_\alpha$ and $b_\alpha$ the conditions (2.2) are essential.

## § 3. Sufficient algebraic conditions

In this paragraph we give some simple algebraic conditions under which the imbeddings

$$W^{\infty}\{a_\alpha, p\} \subset W^{\infty}\{b_\alpha, q\}$$

are valid. More precisely, we get such conditions in two cases:

I. $G = \mathbb{R}^n$, $n \geq 1$, $p = q = 2$;

II. $G = \mathbb{R}^1$, $p = q > 1$ are arbitrary.

In the first case the conditions obtained are necessary and sufficient conditions; in the second case these conditions are only sufficient, however, from our point of view, are close to the necessary conditions.

I. ($G = \mathbb{R}^n$, $n \geq 1$, $p = q = 2$). As it is known from § 2, Chapter I, the nontriviality of the space $W^\infty\{a_\alpha, 2\}$ of functions, the domain of definition of which is the whole space $\mathbb{R}^n$, is defined by the characteristic function

$$a(\xi) = \sum_{|\alpha|=0}^{\infty} a_\alpha \xi^{2\alpha}, \quad \xi = (\xi_1, \ldots, \xi_n).$$

Namely, the space $W^\infty\{a_\alpha, 2\}$ is nontrivial if and only if the function $a(\xi)$ is an analytic function in a neighbourhood of zero. Let us denote by $G_a$ the domain of convergence of the series $a(\xi)$. (We recall that the domain of convergence of $a(\xi)$ is the open kernel of all such points in which this series converges. Some properties of $G_a$ are described, for example, in B. V. ŠABAT [1].) In the same way we denote by $G_b$ the domain of convergence of the characteristic function

$$b(\xi) \equiv \sum_{|\alpha|=0}^{\infty} b_\alpha \xi^{2\alpha}, \quad \xi = (\xi_1, \ldots, \xi_n)$$

of the space $W^\infty\{b_\alpha, 2\}$.

There is the following result (cf. L. R. VOLEVICH and B. P. PANEJAH [1], Theorem 5.1).

Theorem 3.1. The space $W^\infty\{a_\alpha, 2\}$ can be imbedded in the space $W^\infty\{b_\alpha, 2\}$ if and only if the two conditions are satisfied:

1) $G_a \subset G_b$;

2) $\sup_{\xi \in G_a} b(\xi) a^{-1}(\xi) < \infty$.

In this connection

$$\| i_{\infty,\infty} \|^2 = \sup_{\xi \in G_a} b(\xi) a^{-1}(\xi).$$

Proof. For the proof we shall use the following

Lemma 3.1. Let $u(x) \in W^\infty\{a_\alpha, 2\}$. Then the Fourier transform of $u(x)$ is equal to zero for almost all $\xi \bar{\in} G_a$. In addition, the formula

$$\| u \|^2_{a,\infty} = \int_{G_a} a(\xi) |\tilde{u}(\xi)|^2 d\xi \tag{3.1}$$

is valid.

Indeed, the first assertion of this lemma is essentially contained in Lemma 2.1, Chapter I (however, it will be easily proved directly with the help of Parseval's equality and Egorov's theorem). The formula (3.1) is a corollary of Parseval's equality.

Let us turn to the proof of Theorem 3.1.

Sufficiency. Let $u(x) \in W^\infty\{a_\alpha, 2\}$ and let the conditions 1), 2) be fulfilled. Let us show that $u(x) \in W^\infty\{b_\alpha, 2\}$. In fact, from 2) it follows that for any $m = 0,1,\dots$

$$b_m(\xi)a^{-1}(\xi) = \text{const}, \quad \xi \in G_a,$$

where

$$b_m(\xi) \equiv \sum_{|\alpha|=0}^{m} b_\alpha \xi^{2\alpha}.$$

Hence, using Lemma 3.1, we have

$$\int_{G_b} b_m(\xi)|\tilde{u}(\xi)|^2 d\xi = \int_{G_a} b_m(\xi)|\tilde{u}(\xi)|^2 d\xi$$

$$\leq \sup_{\xi\in G_a} b_m(\xi)a^{-1}(\xi) \int_{G_a} a(\xi)|\tilde{u}(\xi)|^2 d\xi.$$

This inequality means that for all $m = 0,1,\dots$ the imbeddings

$$i_{\infty,m}: W^\infty\{a_\alpha, 2\} \to W^m\{b_\alpha, 2\}$$

are valid; moreover,

$$|i_{\infty,m}|^2 \leq \sup_{\xi\in G_a} b_m(\xi)a^{-1}(\xi).$$

The method of contradiction shows that, in fact,

$$|i_{\infty,m}|^2 = \sup_{\xi\in G_a} b_m(\xi)a^{-1}(\xi). \tag{3.2}$$

It is evident, in view of conditions 1), 2) that there exists a finite limit of the norms $|i_{\infty,m}|$, as $m \to \infty$. Consequently, according to Theorem 2.1 (or Theorem 1.1) we get that

$$W^\infty\{a_\alpha, 2\} \subset W^\infty\{b_\alpha, 2\}$$

and

$$\|i_{\infty,\infty}\|^2 = \lim_{m\to\infty}|i_{\infty,m}|^2 \leq \sup_{\xi\in G_a} b(\xi)a^{-1}(\xi). \tag{3.3}$$

Necessity. At first let us prove the necessity of condition 1). Indeed, if condition 1) is not satisfied, then there exists a point $\xi_0 \in G_a$ which, however, does not belong to $G_b$, i. e. $\xi_0 \bar{\in} G_b$. Moreover, there is a neighbourhood $U_{\xi_0}$ of $\xi_0$ which does not intersect $G_b$. Let us consider an arbitrary function $\tilde{u}(\xi) \in C_0^\infty(U_{\xi_0})$. It is evident that the function $u(x) = F^{-1}\tilde{u}(\xi)$ ($F^{-1}$ is the inverse Fourier transform) belongs to the space $W^\infty\{a_\alpha, 2\}$ and, simultaneously, this function does not belong to the space $W^\infty\{b_\alpha, 2\}$, i. e. $u(x) \bar{\in} W^\infty\{b_\alpha, 2\}$.

Thus, if the imbedding

$$W^\infty\{a_\alpha, 2\} \subset W^\infty\{b_\alpha, 2\}$$

is true, then condition 1) is also true.

Let us now prove the necessity of condition 2). Let $W^\infty\{a_\alpha, 2\} \subset W\{b_\alpha, 2\}$. Then, according to Theorem 2.1

$$\lim_{m\to\infty} |i_{\infty,m}| = |i_{\infty,\infty}|,$$

i. e. in view of (3.2)

$$\lim_{m\to\infty} \sup_{\xi\in G_a} b_m(\xi) a^{-1}(\xi) = |i_{\infty,\infty}|^2. \qquad (3.4)$$

Let us choose a sequence of increasing subdomains $G_{a,\varepsilon} \subset G_a$ such that

1) $G_{a,\varepsilon} \to G_a \subset G_b$, as $\varepsilon \to 0$ (condition 1) is already proved);

2) for any subdomain $G_{a,\varepsilon}$

$$b_m(\xi) \to b(\xi)$$

uniformly, as $m \to \infty$.

Then

$$\lim_{m\to\infty} \sup_{\xi\in G_a} \frac{b_m(\xi)}{a(\xi)} \geq \lim_{m\to\infty} \sup_{\xi\in G_{a,\varepsilon}} \frac{b_m(\xi)}{a(\xi)} = \sup_{\xi\in G_{a,\varepsilon}} \frac{b(\xi)}{a(\xi)}.$$

Tending $\varepsilon \to 0$ and taking into account (3.4), we obtain the inequality

$$|i_{\infty,\infty}|^2 \geq \sup_{\xi\in G_a} b(\xi) a^{-1}(\xi), \text{ Q.E.D.} \qquad (3.5)$$

In addition, comparing (3.3) and (3.5) we get that

$$|i_{\infty,\infty}|^2 = \sup_{\xi\in G_a} b(\xi) a^{-1}(\xi).$$

The theorem is completely proved.

II. ($G = \mathbb{R}^1$, $p = q > 1$ arbitrary).

Let

$$W^\infty\{a_n,p\} = \left\{ u(x) \in C^\infty(\mathbb{R}^1) : |u|^p_{a,\infty} = \sum_{n=0}^{\infty} a_n |D^n u|^p_p < \infty \right\}$$

be the Sobolev space of infinite order on the whole real line which is defined by the sequance $a_n$. By analogy, let

$$W^\infty\{b_n,p\} = \left\{ u(x) \in C^\infty(\mathbb{R}^1) : |u|^p_{b,\infty} = \sum_{n=0}^{\infty} b_n |D^n u|^p_p < \infty \right\}$$

be the second space, defined by the sequence $b_n$.

It is clear that the inequalities $b_n \leqq K\,a_n$, where $K > 0$ is a constant, $n = 0,1,\ldots$, guarantee the imbedding $W^\infty\{a_n,p\} \subset W^\infty\{b_n,p\}$. However, these conditions are very limited, since the equalities $a_n = 0$ immediately imply $b_n = 0$ for the corresponding n. The theorem which will be proved below does not have this shortcoming.

Let $R_a$ and $R_b$ denote the radii of convergence of the characteristic functions

$$a(\xi) = \sum_{n=0}^{\infty} a_n \xi^n, \quad b(\xi) = \sum_{n=0}^{\infty} b_n \xi^n, \quad \xi \in \mathbb{R}^1, \tag{3.6}$$

of the spaces $W^\infty\{a_n,p\}$ and $W^\infty\{b_n,p\}$.

<u>Lemma 3.2</u>. Let $u(x) \in W^\infty\{a_n,p\}$. Then for any $n = 0,1,\ldots$ the following inequalities are valid:

$$1) \quad |D^n u|^p_p \leqq 2^p \frac{R_a^n}{\sum_{k=0}^{n} a_k R_a^k} \cdot \sum_{k=0}^{n} a_k |D^k u|^p_p,$$

if $R_a < \infty$ ;

$$2) \quad |D^n u|^p_p = 2^{p+1} \sup_{\xi > 0} \left[\xi^n a^{-1}(\xi)\right] \cdot \sum_{k=0}^{\infty} a_k |D^k u|^p_p,$$

if $R_a = \infty$.

<u>Proof</u>. The known Kolmogorov-Stein inequalities

$$|D^n u|_p \leqq 2 |D^k u|_p^{\frac{m}{n-k+m}} \cdot |D^{n+m} u|_p^{\frac{n-k}{n-k+m}}$$

($k = 0,1,\ldots,n$; $m = 1,2,\ldots$) immediately imply the inequalities

$$|D^n u|^p_p \leqq 2^p \left[\xi^{n-k} |D^k u|^p_p + \xi^{-m} |D^{n+m} u|^p_p\right],$$

where $\xi > 0$ is an arbitrary number.

Hence we get

$$a_k\xi^k a_{n+m}\xi^{n+m}|D^n u|_p^p \leq 2^p\Big[\xi^{n+m}a_{n+m}a_k|D^k u|_p^p$$

$$+ a_k\xi^k a_{n+m}|D^{n+m}u|_p^p\Big]\cdot \xi^n .$$

Summing these inequalities for $k = 0,1,\ldots,m$ and $m = 1,2,\ldots,N$ ($N \geq 1$ is an arbitrary number), we obtain the inequality

$$\left(\sum_{k=0}^{n} a_k\xi^k\right)\left(\sum_{m=1}^{N} a_{n+m}\xi^{n+m}\right)|D^n u|_p^p$$

$$\leq 2^p\xi^n\Big[\sum_{m=1}^{N} a_{n+m}\xi^{n+m}\cdot\sum_{k=0}^{n} a_k|D^k u|_p^p$$

$$+ \sum_{k=0}^{n} a_k\xi^k \sum_{m=1}^{N} a_{n+m}|D^{n+m}u|_p^p\Big].$$

It is equivalent that

$$\sum_{k=0}^{n} a_k\xi^k|D^n u|_p^p \leq 2^p\xi^n\Big[\sum_{k=0}^{n} a_k|D^k u|_p^p$$

$$+ \frac{\sum_{k=0}^{n} a_k\xi^k}{\sum_{m=1}^{N} a_{n+m}\xi^{n+m}}\cdot\sum_{m=1}^{N} a_{n+m}|D^{n+m}u|_p^p\Big]. \tag{3.7}$$

Let us now suppose that $R_a < \infty$. Then we choose $\xi > R_a$ and let $N \to +\infty$ in (3.7). Taking into account the divergence of series $a(\xi)$ (see (3.6)) and the inclusion $u(x) \in W^\infty\{a_n, p\}$, we get the inequality

$$|D^n u|_p^p \leq \frac{2^p\xi^n}{\sum_{k=0}^{n} a_k\xi^k}\cdot\sum_{k=0}^{n} a_k|D^k u|_p^p.$$

Tending $\xi \to R_a$ in the latter inequality we obtain the desired inequality 1) of our lemma.

Let us now consider the case $R_a = \infty$, i. e. the case of an entire characteristic function $a(\xi)$. Let us choose the number $\xi = \xi_n$ so that

$$\sum_{k=0}^{n} a_k\xi_n^k = \sum_{m=1}^{\infty} a_{n+m}\xi_n^{n+m} = \frac{a(\xi_n)}{2}.$$

Then the desired inequality 2) immediately follows from (3.7). Lemma 3.2 is proved.

The next theorem is a direct corollary of Lemma 3.2.

Theorem 3.2. Let the following inequalities be satisfied:

$$1)\ \sum_{n=0}^{\infty} b_n R_a^n \left(\sum_{k=0}^{n} a_k R_a^k\right)^{-1} < \infty , \text{ if } R_a < \infty ;$$

$$2)\ \sum_{n=0}^{\infty} b_n \cdot \sup_{\xi>0} \xi^n a^{-1}(\xi) < \infty , \text{ if } R_a = \infty .$$

Then the imbedding $W^{\infty}\{a_n,p\} \subset W^{\infty}\{b_n,p\}$ holds.

Corollary. Let $R_a < R_b$ or let $R_a = R_b$ and let the point $\xi = R_b$ be a point of convergence of characteristic function $b(\xi)$. Then

$$W^{\infty}\{a_n,p\} \subset W^{\infty}\{b_n,p\} .$$

Remark. The latter corollary shows that the conditions 1), 2) of our theorem are exact enough, since under the condition $R_a > R_b$ there is no imbedding $W^{\infty}\{a_n,p\} \subset W^{\infty}\{b_n,p\}$ by all means.

In conclusion we give some simple examples.

Example 1. Let $a_n = \{1, \text{ if } n = 2k;\ 0, \text{ if } n = 2k+1;\ k = 0,1,\dots\}$. Then $a(\xi) = (1 - \xi^2)^{-1}$ and $R_a = 1$. In this case the condition 1) gives the inequality

$$\sum_{n=0}^{\infty} \frac{b_n}{\left[\frac{n}{2}\right]} < \infty ,$$

the fulfillment of which secures the imdebbing of the space

$$W^{\infty}\{a_n,p\} = \left\{ u(x) \in C^{\infty}(\mathbb{R}^1) :\ \sum_{k=0}^{\infty} \| D^{2k} u \|_p^p < \infty \right\}$$

in the space $W^{\infty}\{b_n,p\}$.

Example 2. Let $a(\xi) = \exp \xi^2$. Then

$$\sup_{\xi>0} \left[\xi^n \exp(-\xi^2)\right] = \left(\frac{n}{2e}\right)^{\frac{n}{2}}$$

and the condition 2) gives the inequality

$$\sum_{n=1}^{\infty} b_n \left(\frac{n}{2e}\right)^{\frac{n}{2}} < \infty ,$$

the fulfillment of which guarantees the imbedding of the space

$$W^{\infty}\{a_n,p\} = \left\{u(x) \in C^{\infty}(\mathbb{R}^1): \sum_{k=0}^{\infty} \frac{1}{k!}|D^{2k}u|_p^p < \infty\right\}$$

in the space $W^{\infty}\{b_n,p\}$.

Further investigations of algebraic conditions for the imbeddings of Sobolev spaces of infinite order were made by G. S. BALASHOVA [2]. In this work a series of easily verified conditions for the imbeddings of the spaces

$$W^{\infty}\{a_n,p,r\} = \left\{u(x) \in C^{\infty}(0,1): \sum_{n=0}^{\infty} a_n|D^n u|_r^p < \infty\right\},$$

where $p \geq 1$, $r \geq 1$, are obtained. The compactness of these imbeddings is also considered. In the case of rapidly decreasing coefficients $a_n$ these conditions are necessary and sufficient conditions.

Let us formulate the main results of BALASHOVA.

Theorem 3.3. Let $a_{n+1} \leq a_n^2$, $a_0 < 1$. The imbedding

$$W^{\infty}\{a_n,p,r\} \subset W^{\infty}\{b_n,p,r\} \tag{3.8}$$

is valid precisely when

$$\overline{\lim} \frac{b_n}{a_n} < \infty.$$

The imbedding (3.8) is compact if and only if

$$\overline{\lim} \frac{b_n}{a_n} = 0.$$

Let us now consider the question of the imbedding

$$\mathring{W}^{\infty}\{a_n,p,r\} \subset W^{\infty}\{b_n,p,r\}, \tag{3.9}$$

where $a_n \geq 0$, $b_n \geq 0$ are arbitrary sequences. Of course, we assume that the corresponding spaces are nontrivial (see § 1, Chapter I).

Theorem 3.4. Let the following condition be fulfilled:

$$\overline{\lim}\, b_n M_n^c v_n < \infty,$$

where $M_n^c$ is the convex regularization of the sequence $a_n^{-1}$ by means of logarithms, $v_n$ being an arbitrary sequence so that

$$\sum_{n=0}^{\infty} v_n^{-1} < \infty .$$

Then there exists the imbedding (3.9); moreover, this imbedding is compact.

If the sequence $a_n^{-1}$ is a logarithmically convex sequence, then the imbedding (3.9) is valid under the condition

$$\overline{\lim}\, b_n\, a_n^{-1} < \infty$$

and this imbedding is compact under the condition

$$\overline{\lim}\, b_n\, a_n^{-1} = 0.$$

The cases of the halfline $\mathbb{R}^1_+$ and the whole line $\mathbb{R}^1$ are also considered.

# CHAPTER VI

# NONSTATIONARY BOUNDARY VALUE PROBLEMS OF INFINITE ORDER

## Introduction

In this Chapter the solvability of the main boundary value problems for parabolic and hyperbolic equations of infinite order is established. Besides, the Schrödinger equation of infinite order is also considered.

The method of the study of these problems is, principally, just the same, as in the stationary case:

1) solving of the approximate equations of order 2m;

2) pass to the limit, as $m \to \infty$.

In this connection we are concerned (for sake of brevity) only with the consideration of model equations:

$$\frac{\partial u}{\partial t} + L(u) = h(t,x),$$

$$\frac{\partial^2 u}{\partial t^2} + L(u) = h(t,x),$$

$$\frac{1}{i}\frac{\partial u}{\partial t} + L(u) = h(t,x),$$

where

$$L(u) \equiv \sum_{|\alpha|=0}^{\infty} (-1)^{|\alpha|} D^\alpha(a_\alpha |D^\alpha u|^{p-2} D^\alpha u)$$

is a nonlinear elliptic operator of infinite order. In the same way the general case may be considered.

## § 1. First boundary value problem for parabolic equations of infinite order

Let $G \subset \mathbb{R}^n$ be a bounded domain, $Q = [0,T] \times G$ be a cylinder with lateral surface $S = [0,T] \times \Gamma$, where $\Gamma$ is the boundary of $G$.

In the cylinder Q we study the first boundary value problem

$$\frac{\partial u}{\partial t} + \sum_{|\alpha|=0}^{\infty} (-1)^{|\alpha|} D^\alpha(a_\alpha |D^\alpha u|^{p-2} D^\alpha u) = h(t,x), \tag{1.1}$$

$$u(o,x) = 0 \tag{1.2}$$

$$D^{\omega}u|_S = 0, \quad |\omega| = 0,1,\ldots \tag{1.3}$$

Here $a_\alpha \geq 0$ is a sequence of real numbers, $p > 1$ is a number. The "energy" space

$$\mathring{W}^{\infty}\{a_\alpha, p\} = \left\{ u(x) \in C_0^{\infty}(G) : \|u\|_{\infty}^{p} = \sum_{|\alpha|=0}^{\infty} a_\alpha \|D^\alpha u\|_p^p < \infty \right\}$$

of function $u(x): G \longrightarrow \mathbb{C}^1$ corresponds to the elliptic part of equation (1.1).

Our main assumption is that this space is nontrivial (see § 1, Chapter I).

Let us denote by $L_p(0,T;\mathring{W}^{\infty}\{a_\alpha,p\})$ the space of functions $u(t,x)$ which have finite norm

$$\|u\|_{p,\infty}^{p} = \int_0^T \|u\|_{\infty}^{p}\, dt$$

and are equal to zero together with all derivatives $D^{\omega}u$ on the lateral surface S.

In other words

$$L_p(0,T;\mathring{W}^{\infty}\{a_\alpha,p\})$$

$$= \left\{ u(t,x) : \sum_{|\alpha|=0}^{\infty} a_\alpha \int_0^T \|D^\alpha u\|_p^p\, dt < \infty\ ,\ D^{\omega}u|_S = 0,\ |\omega| = 0,1,\ldots \right\}.$$

Further, let $L_{p'}(0,T;W^{-\infty}\{a_\alpha,p'\})$ be the dual space of the space $L_p(0,T;\mathring{W}^{\infty}\{a_\alpha,p\})$ that is the space of generalized functions $h(t,x)$ having a form

$$h(t,x) = \sum_{|\alpha|=0}^{\infty} (-1)^{|\alpha|} a_\alpha D^\alpha h_\alpha(t,x),$$

where $h_\alpha(t,x) \in L_{p'}(Q)$ and

$$\sum_{|\alpha|=0}^{\infty} a_\alpha \int_0^T \|h_\alpha(t,x)\|_{p'}^{p'}\, dt < \infty\ . \tag{1.4}$$

The value of $h(t,x) \in L_{p'}(0,T;W^{-\infty}\{a_\alpha,p'\})$ on an element $v(t,x) \in L_p(0,T;\mathring{W}^{\infty}\{a_\alpha,p\})$ is defined by formula

$$\int_0^T \langle h,v\rangle\, dt \equiv \sum_{|\alpha|=0} a_\alpha \int_0^T \int_G h_\alpha(t,x) D^\alpha v(t,x)\, dx dt,$$

which, as it easy to see, is correct.

By definition, two functions $h_1(t,x)$ and $h_2(t,x)$ from the space $L_{p'}(0,T;W^{-\infty}\{a_\alpha,p'\})$ are equal if for any function $v(t,x) \in L_p(0,T;\mathring{W}^{\infty}\{a_\alpha,p\})$ there is equality

$$\int_0^T \langle h_1,v\rangle\, dt = \int_0^T \langle h_2,v\rangle\, dt.$$

The space $L_{p'}(0,T;W^{-\infty}\{a_\alpha,p'\})$ may be considered as a Banach space with norm

$$\|h\|_{p',-\infty} \equiv \sup \frac{\int_0^\infty \langle h,v\rangle\, dt}{\|v\|_{p,\infty}}, \quad v(t,x) \in L_p(0,T;\mathring{W}^{\infty}\{a_\alpha,p\}).$$

<u>Theorem 1.1.</u> For any right side $h(t,x) \in L_{p'}(0,T;W^{-\infty}\{a_\alpha,p'\})$ there exists one and only one function $u(t,x)$, satisfying the following conditions

1) $u \in L_p(0,T;\mathring{W}^{\infty}\{a_\alpha,p\})$, $\frac{\partial u}{\partial t} \in L_{p'}(0,T;W^{-\infty}\{a_\alpha,p'\})$;

2) $u(0,x) = 0$;

3) for any function $v(t,x) \in L_p(0,T;\mathring{W}^{\infty}\{a_\alpha,p\})$

the following identity

$$\int_0^T \langle \frac{\partial u}{\partial t},v\rangle\, dt + \sum_{|\alpha|=0}^{\infty} a_\alpha \int_0^T \int_G |D^\alpha u|^{p-2} D^\alpha u D^\alpha v\, dxdt = \int_0^T \langle h,v\rangle\, dt \quad (*)$$

is valid.

<u>Proof</u>. The proof of Theorem 1.1 is parallel to the proof of the solvability of the Dirichlet problem for an elliptic equation, given in Chapter II, § 1, and it differs from the latter in only one step (namely, in the choice of a convergent subsequence of solutions).

Thus, let $u_m(t,x)$, $m = 1,2,\ldots$, be the family of solutions of boundary value problems

$$\frac{\partial u_m}{\partial t} + \sum_{|\alpha|=0} (-1)^{|\alpha|} D^\alpha(a_\alpha |D^\alpha u_m|^{p-2} D^\alpha u_m) = h_m(t,x); \qquad (1.4)_m$$

$$u_m(0,x) = 0 \qquad (1.5)_m$$

$$D^\omega u_m|_S = 0, \quad |\omega| \leq m-1, \qquad (1.6)_m$$

where

$$h_m(t,x) = \sum_{|\alpha|=0}^{m} (-1)^{|\alpha|} a_\alpha D^\alpha h_\alpha(t,x).$$

The solvability of the problems $(1.4)_m$ - $(1.6)_m$ is well known (see M. I. VIŠIK [1], F. E. BROWDER [2], J. L. LIONS [1]); moreover, there is the estimate

$$\sum_{|\alpha|=0}^{m} a_\alpha \int_0^T \|D^\alpha u_m\|_p^p \, dt \leqq K, \tag{1.7}$$

where $K > 0$ depends on the value (1.4) but does not depend on $m = 1,2,\ldots$

Our first aim is to choose a subsequence of the sequence of solutions $u_m(t,x)$, which converges in $L_p(Q)$ strongly (with all its derivatives of type $D^{\omega}$). For this we remark that from estimate (1.7) and equation $(1.4)_m$ it immediately follows that

$$\frac{\partial u_m}{\partial t} \in L_{p'}(0,T;W_{p'}^{-m}(G)),$$

where $W_{p'}^{-m}(G)$ is the dual space of the Sobolev space $\mathring{W}_p^m(G)$, and, a fortiori,

$$\frac{\partial u_m}{\partial t} \in L_{p'}(0,T;W^{-\infty}\{a_\alpha,p'\}).$$

In addition, for any function $v(t,x) \in L_p(0,T;\mathring{W}^\infty\{a_\alpha,p\})$ the following inequality is valid:

$$\int_0^T \left|\left\langle \frac{\partial u_m}{\partial t}, v\right\rangle\right| dt \leqq \sum_{|\alpha|=0}^{m} a_\alpha \int_0^T \int_G |h_\alpha D^\alpha v| \, dxdt$$

$$+ \sum_{|\alpha|=0}^{m} a_\alpha \int_0^T \int_G |D^\alpha u_m|^{p-1} \cdot |D^\alpha v| \, dxdt$$

$$\leqq \left[\left(\sum_{|\alpha|=0}^{m} a_\alpha \int_0^T \|h_\alpha\|_{p'}^{p'} \, dt\right)^{\frac{1}{p'}} + \left(\sum_{|\alpha|=0}^{m} a_\alpha \int_0^T \|D^\alpha u_m\|_p^p \, dt\right)^{\frac{1}{p'}}\right] \cdot$$

$$\cdot \left(\sum_{|\alpha|=0}^{m} a_\alpha \int_0^T \|D^\alpha v\|_p^p \, dt\right)^{\frac{1}{p}} \leqq \left([\varrho'(h)]^{\frac{1}{p'}} + K^{\frac{1}{p'}}\right) \|v\|_{p,\infty}.$$

Hence is follows that

$$\left\|\frac{\partial u_m}{\partial t}\right\|_{p',-\infty} \leqq \left([\varrho'(h)]^{\frac{1}{p'}} + K^{\frac{1}{p'}}\right), \tag{1.8}$$

i. e. the derivatives $\partial u_m/\partial t$ form a bounded set in the space $L_{p'}(0,T;W^{-\infty}\{a_\alpha,p'\})$.

The estimates (1.7) and (1.8) permit us to apply the well-known (in the theory of nonstationary problems) lemma of compactness (see some variants of this lemma in J. L. LIONS [2], J. AUBIN [1],

JU. A. DUBINSKIJ [1]). Let's recall the formulation of this lemma. Let $B_0$, $B$, $B_1$ be Banach spaces. Let us put

$$Y = \left\{u(t):\ u(t) \in L_{p_0}(0,T;B_0),\ u'(t) \in L_{p_1}(0,T;B_1)\right\},$$

where $p_0 > 1$, $p_1 > 1$ are numbers.

Lemma (of compactness). Let the imbeddings

$$B_0 \subset B \subset B_1$$

hold; moreover, let the imbedding $B_0 \subset B$ be compact. Then

$$Y \subset L_{p_0}(0,T;B)$$

and this imbedding is compact.

In order to apply this lemma we put:

$$B_0 \equiv W^{N+1}\left\{a_\alpha,p\right\} \overset{\text{def}}{\equiv} \left\{u(x):\ \sum_{|\alpha|=0}^{N} a_\alpha \|D^\alpha u\|_p^p < \infty\right\};$$

$$B \equiv W^N\left\{a_\alpha,p\right\};\ B_1 \equiv W^{-\infty}\left\{a_\alpha,p'\right\};\quad p_0 = p;\ p_1 = p'$$

($p' = p/(p-1)$), where $N \geq 0$ is arbitrary.

Then, in view of estimates (1.7) and (1.8) we get that the family $u_m(t,x)$ of solutions of the problems $(1.4)_m$ - $(1.6)_m$ is compact in the space $L_p(0,T;W^N\{a_\alpha,p\})$, where $N$ is arbitrary. Consequently, one can assume (using the diagonal process, if needed) that the sequence $u_m(t,x)$ converges together with all derivatives $D^\omega u_m(t,x)$ to a function $u(t,x) \in L_p(0,T;\overset{\circ}{W}{}^\infty\{a_\alpha,p\})$ in the space $L_p(Q)$ strongly.

Passing to the limit in $(1.4)_m$ - $(1.6)_m$, as $m \to \infty$, we obtain (in the same way, as in the elliptic case) that the limit function $u(t,x)$ satisfies the identity (*). In addition, $u(0,x) = 0$. Thereby, $u(t,x)$ is the desired solution of the original problem (1.1) - (1.3).

The monotonicity of the operator $L(u)$ implies, as usual, the uniqueness of our solution.

The theorem is proved.

## § 2. Mixed problems for hyperbolic equations of infinite order

Let, as in § 1, $Q$ be a cylinder with the lateral surface $S$, i. e. $Q = [0,T] \times G$, where $G \subset \mathbb{R}^n$ is a bounded domain. In this cylinder we consider the mixed problem of infinite order

$$\frac{\partial^2 u}{\partial t^2} + \sum_{|\alpha|=0} (-1)^{|\alpha|} D^\alpha(a_\alpha |D^\alpha u|^{p-2} D^\alpha u) = h(t,x) \tag{2.1}$$

$$u(0,x) = 0, \; \frac{\partial u}{\partial t}(0,x) = 0 \tag{2.2}$$

$$D^\omega u|_S = 0, \quad |\omega| = 0,1,\ldots, \tag{2.3}$$

where $a_\alpha \geq 0$ is a sequence of numbers, $p \geq 2$ is a number (the case $1 < p < 2$ requires some slight modifications).

We shall assume that the space $\mathring{W}^\infty\{a_\alpha, p\}$ of functions $u(x): G \longrightarrow \mathbb{C}^1$ is nontrivial. Further, let the right side $h(t,x)$ be a function having the form

$$h(t,x) = \sum_{|\alpha|=0}^{\infty} (-1)^{|\alpha|} a_\alpha D^\alpha h_\alpha(t,x),$$

where for almost all $t \in (0,T)$ the functions $h_\alpha(t,x) \in L_{p'}(G)$ $(p' = p/(p-1)$, $h'_\alpha(t,x) \in L_{p'}(Q)$; moreover,

$$\sup_t \sum_{|\alpha|=0}^{\infty} a_\alpha \|h_\alpha\|_{p'}^{p'} < \infty, \quad \sum_{|\alpha|=0}^{\infty} a_\alpha \int_0^T \|h'_\alpha\|_{p'}^{p'}(t)\,dt < \infty. \tag{2.4}$$

There exists the following result about the solvability in the sense of generalized functions of the problem just described.

Theorem 2.1. For any function $h(t,x)$, mentioned above, there exists at least one function $u(t,x)$, satisfying the following conditions:

1) $u(t,x)$ has the finite integral of "energy"

$$I \equiv \sup_t \left\|\frac{\partial u}{\partial t}\right\|_2^2 + \sup_t \sum_{|\alpha|=0}^{\infty} a_\alpha \|D^\alpha u\|_p^p;$$

2) the following identity is valid:

$$-\int_0^T \left\langle \frac{\partial u}{\partial t}, \frac{\partial v}{\partial t} \right\rangle dt + \sum_{|\alpha|=0}^{\infty} a_\alpha \int_0^T \langle |D^\alpha u|^{p-2} D^\alpha u, D^\alpha v\rangle\, dt = \int_0^T \langle h,v\rangle\, dt, \quad (*)$$

where $v(t,x)$ is an arbitrary function such that

$$\int_0^T \left\|\frac{\partial v}{\partial t}\right\|^2 dt + \sum_{|\alpha|=0}^{\infty} a_\alpha \int_0^T \|D^\alpha v\|_p^p \, dt < \infty;$$

moreover, $v(T,x) = 0$, $D^\omega v|_S = 0$, $|\omega| = 0,1,\ldots$

3) the conditions (2.3) hold (the equality $\frac{\partial u}{\partial t}(0,x) = 0$ holds in the sense of distributions on the basic space $\mathring{W}^{\infty}\{a_\alpha,p\}$, i. e. for any function $w(x) \in \mathring{W}^{\infty}\{a_\alpha,p\}$

$$\lim_{t\to 0} \left\langle \frac{\partial u}{\partial t}(t,x), w(x)\right\rangle = 0).$$

Proof. Let us consider the family of mixed problems of finite order $2m+2$ $(m = 0,1,\ldots)$:

$$\frac{\partial^2 u_m}{\partial t^2} + \sum_{|\alpha|=m+1} (-1)^{m+1} a_\alpha D^{2\alpha} u_m + L_{2m}(u_m) = \tilde{h}_m(t,x), \qquad (2.5)_m$$

$$u_m(0,x) = 0, \quad \frac{\partial u_m}{\partial t}(0,x) = 0, \qquad (2.6)_m$$

$$D^{\omega} u_m\big|_S = 0, \quad |\omega| = 0,1,\ldots, \qquad (2.7)_m$$

where

$$L_{2m}(u_m) \equiv \sum_{|\alpha|=0}^{m} (-1)^m D^\alpha (a_\alpha |D^\alpha u_m|^{p-2} D^\alpha u_m)$$

$$\tilde{h}_m(t,x) = \sum_{|\alpha|=0}^{m} (-1)^{|\alpha|} a_\alpha D^\alpha h_\alpha(t,x).$$

The problem $(2.5)_m$ - $(2.7)_m$ is a weakly nonlinear hyperbolic problem. Therefore, it has at least one generalized solution $u_m(t,x)$. This solution may be obtained, for example, by method of Galerkin-Faedo-Hopf (see J. L. LIONS [3]); in addition,

$$\sup_t \left[ \left\| \frac{\partial u_m}{\partial t} \right\|_2^2 + \sum_{|\alpha|=m+1} a_\alpha \| D^\alpha u_m \|_2^2 + \sum_{|\alpha|=0}^{m} a_\alpha \| D^\alpha u_m \|_p^p \right] \leq K, \quad (2.8)$$

where the constant $K > 0$ does not depend on $m = 1,2,\ldots$ (In the proof of estimate (2.8), which is done by standard calculations, the conditions (2.4) on function $h(t,x)$ are taken into account.)

From estimate (2.8) and the imbedding theorems of Sobolev anisotropic spaces it follows that there exists a function $u(t,x)$, having the finite integral of "energy" I (see condition 1) of our theorem) and a subsequence of the sequence $u_m(t,x)$ such that

$$D^\alpha u_m \to D^\alpha u \text{ in } L_p(Q) \text{ strongly } (|\alpha| = 0,1,\ldots);$$

$$\frac{\partial u_m}{\partial t} \to \frac{\partial u}{\partial t} \quad \text{in } L_2(Q) \text{ weakly}.$$

In the same way, as in § 2, Chapter II (see the passage to the limit in the family of problems $(2.3)_m$, $(2.4)_m$, as $m \to \infty$), we get that the function u(t,x) satisfies the condition 2) of the theorem.

It is clear that $u(0,x) = 0$, $D^{\omega}u|_S = 0$, $|\omega| = 0,1,\ldots$. It remains to show that $\partial u/\partial t(0,x) = 0$ too. For this let us put $v(t,x) = \varphi(t)\psi(x)$, where $\varphi(t) \in C_0^{\infty}(0,T)$ and $\psi(x) \in \mathring{W}^{\infty}\{a_\alpha, p\}$, in the identity (*). Then, integrating by parts, we get the identity

$$\int_0^T \langle u, \psi(x)\rangle \varphi''(t)dt + \sum_{|\alpha|=0} a_\alpha \int_0^T \langle |D^\alpha u|^{p-2} D^\alpha u, D^\alpha \psi(x)\rangle \varphi(t)dt$$

$$= \int_0^T \langle h, \psi(x)\rangle \varphi(t)dt,$$

which means that the solution u(t,x) satisfies the equation

$$\frac{\partial^2 u}{\partial t^2} + \sum_{|\alpha|=0}^{\infty} (-1)^{|\alpha|} D^\alpha(a_\alpha |D^\alpha u|^{p-2} D^\alpha u) = h(t,x) \qquad (2.9)$$

in the sense of distributions on the basic space $C_0^{\infty}(0,T;\mathring{W}^{\infty}\{a_\alpha,p\})$ (the notation is clear).

Since $u(t,x) \in L_p(0,T;\mathring{W}^{\infty}\{a_\alpha,p\})$ and $h(t,x) \in L_{p'}(0,T;W^{-\infty}\{a_\alpha,p'\})$, then from (2.9) we get that $\partial^2 u/\partial t^2 \in L_{p'}(0,T;W^{-\infty}\{a_\alpha,p'\})$ too. Consequently, the equation (2.9) holds in the sense of distributions on the basic space $L_p(0,T;\mathring{W}^{\infty}\{a_\alpha,p\})$, i. e.

$$\int_0^T \langle u'', v\rangle\, dt + \sum_{|\alpha|=0}^{\infty} a_\alpha \int_0^T \langle |D^\alpha u|^{p-2} D^\alpha u, D^\alpha v\rangle\, dt$$

$$= \int_0^T \langle h, v\rangle\, dt \qquad (2.10)$$

for any function $v(t,x) \in L_p(0,T;\mathring{W}^{\infty}\{a_\alpha,p\})$. (We emphasize that no conditions of the basic functions v(t,x) on $t \in (0,T)$ or conditions of type $v(0,x) = 0$ or $v(T,x) = 0$ are required.)

In particular, let us put in (2.10) an arbitrary function v(t,x), satisfying condition 2) of the theorem, i. e. an arbitrary function for which

$$\int_0^T \left\|\frac{\partial v}{\partial t}\right\|_2^2 dt + \sum_{|\alpha|=0}^{\infty} a_\alpha \int_0^T \|D^\alpha v\|_p^p\, dt < \infty$$

and $v(T,x) = 0$, $D^{\omega}v|_S = 0$, $|\omega| = 0,1,\ldots$

Then, integrating in (2.10) by parts, we get that

$$-\int_0^T \left\langle \frac{\partial u}{\partial t}, \frac{\partial v}{\partial t}\right\rangle dt + \left\langle \frac{\partial u}{\partial t}, v\right\rangle\Big|_{t=0} + \sum_{|\alpha|=0}^{\infty} a_\alpha \int_0^T \langle |D^\alpha u|^{p-2} D^\alpha u, D^\alpha v\rangle\, dt$$

$$= \int_0^T \langle h,v\rangle \, dt.$$

Comparing the latter identity with the main identity (*), we immediately obtain that

$$\left\langle \frac{\partial u}{\partial t}, v\right\rangle \Big|_{t=0} = 0.$$

Therefore, $\partial u/\partial t(0,x) = 0$, since the values $v(0,x)$ are arbitrary. The theorem is proved.

## § 3. Mixed problems for the nonlinear Schrödinger equation of infinite order

In the cylinder $Q = [0,T] \times G$ the following problem is considered:

$$\frac{1}{i}\frac{\partial u}{\partial t} + \sum_{|\alpha|=0}^{\infty} (-1)^{|\alpha|} D^{\alpha}(a_{\alpha}|D^{\alpha}u|^{p-2} D^{\alpha}u) = h(t,x), \tag{3.1}$$

$$u(0,x) = 0, \tag{3.2}$$

$$D^{\omega}u\big|_S = 0, \quad |\omega| = 0,1,\dots \tag{3.3}$$

Let us suppose that for the elliptic part of the equation (3.1) and for the right side, all conditions in § 2 are fulfilled. Under these conditions there is the following

Theorem 3.1. For any function $h(t,x)$ there exists at least one function $u(t,x)$ satisfying the following conditions:

1) $\sup\limits_t \sum\limits_{|\alpha|=0}^{\infty} a_{\alpha}\|D^{\alpha}u\|_p^p < \infty$, $D^{\omega}u\big|_S = 0$, $|\omega| = 0,1,\dots;$

2) $\frac{\partial u}{\partial t} \in L_{p'}(0,T;W^{-\infty}\{a_{\alpha},p'\})$, $u(0,x) = 0;$

3) for any function $v(t,x) \in L_p(0,T;\mathring{W}^{\infty}\{a_{\alpha},p\})$ the identity

$$\int_0^T \left\langle \frac{\partial u}{\partial t}, v\right\rangle dt + \sum_{|\alpha|=0}^{\infty} a_{\alpha}\int_0^T \langle |D^{\alpha}u|^{p-2} D^{\alpha}u, D^{\alpha}v\rangle \, dt = \int_0^T \langle h,v\rangle \, dt \tag{3.4}$$

is valid.

Proof. The proof is analogous to those proofs which were done in § 1 and § 2, so we shall be brief.

Let $u_m(t,x)$, $m = 1,2,\dots$, be the solutions of the approximate problems

$$\frac{1}{i}\frac{\partial u_m}{\partial t} + \sum_{|\alpha|=m+1} (-1)^{m+1} a_\alpha D^\alpha u_m$$

$$+ \sum_{|\alpha|=0}^{m} (-1)^{|\alpha|} D^\alpha (a_\alpha |D^\alpha u_m|^{p-2} D^\alpha u_m) = h_m(t,x), \qquad (3.5)_m$$

$$u_m(0,x) = 0, \qquad (3.6)_m$$

$$D^\omega u_m\big|_S = 0, \quad |\omega| \leq m, \qquad (3.7)_m$$

where

$$h_m(t,x) = \sum_{|\alpha|=0}^{m} (-1)^{|\alpha|} a_\alpha D^\alpha h_\alpha(t,x).$$

The problems $(3.5)_m$ - $(3.7)_m$ are solvable as weakly nonlinear problems (the solutions may be obtained, for example, by Galerkin-Faedo-Hopf method); in addition,

$$\sup_t \left[ \sum_{|\alpha|=m+1} a_\alpha \| D^\alpha u_m \|_2^2 + \sum_{|\alpha|=0}^{m} a_\alpha \| D^\alpha u_m \|_p^p \right] \leq K,$$

where the constant $K > 0$ depends on $h(t,x)$ but does not depend on $m = 1,2,\dots$

From this estimate and equation $(3.5)_m$ it follows that the derivatives

$$\frac{\partial u_m}{\partial t} \in L_{p'}(0,T; W^{-\infty}\{a_\alpha, p'\})$$

and form a bounded set in this space. Consequently, applying the "compactness" lemma from § 1, we can assume that a subsequence of the sequence $u_m(t,x)$ converges with all its derivatives to a function $u(t,x) \in L_p(0,T; \mathring{W}^\infty\{a_\alpha, p\})$ in $L_p(Q)$ strongly. As in §§ 1,2, we find that the function $u(t,x)$ satisfies the identity (3.4), i. e. this function is the desired solution. The theorem is proved.

Remark. The operator

$$L(u) \equiv \sum_{|\alpha|=0}^{\infty} (-1)^{|\alpha|} D^\alpha (a_\alpha |D^\alpha u|^{p-2} D^\alpha u)$$

is a model operator of infinite order. The theory of the present Chapter is also valid for the general operator

$$\sum_{|\alpha|=0}^{\infty} (-1)^{|\alpha|} D^\alpha A_\alpha(x, D^\gamma u), \quad |\gamma| \leq |\alpha|,$$

where $A_\alpha(x,\xi_\gamma) \equiv \partial F(x,\xi_\gamma)/\partial \xi_\alpha$ are nonlinear functions of polynomial growth (see § 2, Chapter II and JU. A. DUBINSKIJ [3]).

We also note that, if there exists a "trace" theory in the space $W^\infty\{a_\alpha,p\}$, the present method extends to the proof of the solvability of nonhomogeneous boundary problems.

## APPENDIX

The Sobolev spaces of infinite order have extensive applications to the theory of partial differential equations and, among their number, in mathematical physics. The basis of these applications is the nonformal algebra of differential operators of infinite orders as the operators, acting in the corresponding Sobolev spaces of infinite order. This makes it possible, by considering $\partial/\partial x$ as a parameter, to solve a partial differential equation as an ordinary differential equation, to which are adjoined the initial or boundary conditions.

Considering a partial differential equation of finite order in the space $W^{\infty}\{a_\alpha, 2\}$ we do not require that this equation have any definite type. The type of the equation has no role. We emphasize, however, that in the process of solving of problems that are well-posed in spaces of finite smoothness, the spaces $W^{\infty}\{a_\alpha, 2\}$ play in intermediate role, being merely an instrument of the investigation. But in the solution of problems, that are ill-posed in the sense of Hadamard-Petrovskij, the introduction of the space $W^{\infty}\{a_\alpha, 2\}$ constitutes the very essence of the approach. A series of examples in § 4, P. II is given (for the general case see JU. A. DUBINSKIJ [9] - [12]). The spaces $W^{\infty}\{a_\alpha, 2\}$ which arise in these examples are closely connected with the corresponding concrete problem (in particular, with the Cauchy problem). One may introduce such space of type $W^{\infty}\{a_\alpha, 2\}$ for which any Cauchy problem will be correct. Namely, let us

$$H^{\infty}(\mathbb{R}^n) \overset{\text{def}}{=} \lim_{a(\xi)>0} \operatorname{proj} H^{\infty}_a(\mathbb{R}^n),$$

where

$$a(\xi) \equiv \sum_{|\alpha|=0}^{\infty} a_\alpha \xi^{2\alpha}, \quad a_\alpha \geq 0,$$

is any entire function (characteristic function of the Sobolev space of infinite order $H^{\infty}_a(\mathbb{R}^n)$).

One can show that $H^{\infty}(\mathbb{R}^n)$ is the subspace of the $L_2(\mathbb{R}^n)$, consisting of such functions $u(x) \in L_2(\mathbb{R}^n)$ the Fourier transforme of which are finite. The last property of $H^{\infty}(\mathbb{R}^n)$ permits, for instance, giving a positive answer to the question of correctness of the Cauchy problem for any partial differential equation with constant coefficients (JU. A. DUBINSKIJ [12]). Namely, let us consider the Cauchy problem ($m \geq 1$)

$$\frac{\partial^m u}{\partial t^m} + \sum_{k=0}^{m-1} A_k(t,D) \frac{\partial^k u}{\partial t^k} = h(t,x), \quad t \in \mathbb{R}^1, \quad x \in \mathbb{R}^n, \tag{1}$$

$$\frac{\partial^k u}{\partial t^k}(0,x) = \varphi_k(x), \quad k = 0,1,\ldots,m-1, \tag{2}$$

where $A_k(t,D)$ are arbitrary differential operators (finite or infinite order) with continious coefficients.

Theorem A.1. For all functions $h(t,x) \in C(\mathbb{R}^1, H^\infty(\mathbb{R}^n))$, $\varphi_k(x) \in H^\infty(\mathbb{R}^n)$ there is only one solution $u(t,x) \in C^m(\mathbb{R}^1, H^\infty(\mathbb{R}^n))$ of the problem (1), (2).

Further applications are connected with the theory of the pseudo-differential operators the symbols of which are analytic functions.

Let us mark the result of K. L. SAMAROV [1], where the formula for the fundamental solution of the Cauchy problem for the Schrödinger equation

$$i\frac{\partial u}{\partial t} = \frac{c}{\omega}\sqrt{I - \omega^2 \Delta}\, u, \quad t \in \mathbb{R}^1, \quad x \in \mathbb{R}^3, \tag{3}$$

$$u(0,x) = \varphi(x) \tag{4}$$

(see example 8, § 4, P. II) is obtained in the following form.

Theorem A.2. Fundamental solution of problem (3), (4) is

$$E(t,x) = \begin{cases} -\frac{1}{4\pi c|x|} \frac{\partial}{\partial r} \frac{\partial}{\partial t} H_0^2(\operatorname{sgn} t \frac{1}{\omega}\sqrt{c^2t^2 - r^2}), & |ct| > r; \\ \frac{1}{2\pi c|x|} \frac{\partial}{\partial r} \frac{\partial}{\partial t} K_0^2(\frac{1}{\omega}\sqrt{r^2 - c^2t^2}, & |ct| < r. \end{cases}$$

Here $H_0^2$ is the Hankel function of order zero; $K_0^2$ is the modified Bessel function of order zero; $r = |x|$.

The abstract Sobolev spaces of infinite order were considered by S. R. UMAROV [1]. Let H be a complex Hilbert space and $L : D(L) \to H$ be some normal operator. Let us consider the space

$$W_H^\infty\{L, a_n, p_n\} = \left\{u \in \bigcap_{n=0}^{\infty} D(L^n) : \sum_{n=0}^{\infty} a_n \|L^n u\|^{p_n} < \infty\right\},$$

where $a_n \geq 0$; $p_n \geq 1$ are constants.

Notations: $\sigma = \sigma(\sqrt{L^*L})$ is the spectrum of the operator $\sqrt{L^*L}$; $\sigma' \subset \sigma$ is the subset consisting of the limit points of $\sigma$ and of the eigenvalues of $\sqrt{L^*L}$ of infinite multiplicity; $\sigma^* = \inf \sigma'$.

Theorem A.3. The space $W_H^\infty\{L,a_n,p_n\}$ is nontrivial if and only if the following conditions are satisfied:

1) if there exists the sequence $\sigma_n \in \sigma(\sqrt{L^*L})$ such that $\sigma_n \uparrow \sigma_*$, then

$$\sum_{n=0}^{\infty} a_n q^{np_n} < \infty$$

for all $q < \sigma^*$;

2) if $\sigma_*$ is an eigenvalue of infinite multiplicity, then

$$\sum_{n=0}^{\infty} a_n \sigma_*^{np_n} < \infty ;$$

3) for all other cases there exists $q_1 > \sigma_*$ such that

$$\sum_{n=0}^{\infty} a_n q_1^{np_n} < \infty .$$

The various results in the theory of differential equations of infinite order with arbitrary nonlinearity were obtained by CHAN DYK VAN. These results are published in his doctoral dissertation [6].

Let us describe the last results of G. S. BALASHOVA [3] - [6].

Imbedding theorem. The spaces $W^\infty\{a_n,p,r\}$ of periodic functions $u(x): \mathbb{R}^1 \to \mathbb{R}^1$ are considered.

Theorem A.4. Let one of the following conditions be satisfied:

a) $\overline{\lim\limits_{n\to\infty}}\, b_n a_n^{-1} < \infty$;

b) there exists a sequence $v_n \downarrow 0$ such that

$$\sum_{n=0}^{\infty} v_n < \infty , \quad \overline{\lim_{n\to\infty}}\, b_n (a_n^{-1})^c v_n^{-1} < \infty ;$$

c) the sequence $a_n^{-1}$ is almost logarithmically convex and

$$\lim_{n\to\infty} b_n (a_n^{-1})^c = 0;$$

d) if $p = r > 1$, then

$$\sum_{n=0}^{\infty} b_n \sup_{\xi \geq 0} \left[\xi^n a^{-1}(\xi)\right] < \infty ,$$

where

$$a(\xi) \equiv \sum_{n=0}^{\infty} a_n \xi^n .$$

Then

1) $W^{\infty}\{a_n,p,r\} \subset W^{\infty}\{b_n,p,r\}$;

2) under conditions b), c) and d) the last imbedding is compact.

The method of the proof gives a more precise variant of the known result of S. Mandelbrojt.

Theorem A.5. Let $A_n > 0$, $M_n > 0$ be sequences with $M_n$ a logarithmically convex sequence. Then the condition $\lim_{n\to\infty} A_n M_n^{-1} = 0$ implies that $C(A_n) \subset C(M_n)$ and $C(M_n) \setminus C(A_n) \neq \emptyset$. (Here $C(M_n)$ is Hadamard's class, that is $C(M_n) = \{u(x) \in C^{\infty}(S^1):\ \max|D^n u(x)| \leq K^n M_n$, where $K = K(u)$ is any constant$\}$.)

Trace theorem A.6. Let one of the following conditions hold:

1) there exists a $\lambda \in \mathbb{R}^1$ such that for the given sequence $\{b_m\}$, $m = 0,1,\ldots$,

$$\sum_{m=0}^{\infty} |b_m| \min_{0\leq k\leq m-1} \max \left\{ \frac{(3s_{m-k})^{k+1}\lambda^{m-k-1}}{(k+1)!\,(m-k-1)!}, \right.$$

$$\left. \sum_{i=0}^{k} \frac{(3s_{m-k})^{k-i}}{(k-i)!\, M^c_{m-k+i}} \right\} < \infty ,$$

where

$$s_m = \sum_{n=m}^{\infty} M_n^c (M_{n+1}^c)^{-1};$$

2) the sequence $\left\{a_n^{-\frac{1}{p}}\right\}$ is logarithmically onvex and there exists $\alpha > 1$ such that

$$|b_n| < a_n^{-\frac{1}{p}} n^{-n\alpha};$$

3) the sequence $\left\{a_n^{-\frac{1}{p}} n^{-n\alpha}\right\}$ is logarithmically convex and $\lim_{n\to\infty} (b_n a_n^{1/p})^{1/n} = 0$.

Then there exists a function $f(x) \in W^{\infty}\{a_n,p,r\}\ (0,a)$ such that $D^n f(0) = b_n$, $n = 0,1,\ldots$

These results are applied to the Dirichlet problem of infinite order.

## BIBLIOGRAPHY


ARONSZAJN, N.

[1] Boundary value problems for polyharmonic functions, preprint.

AUBIN, J.

[1] Un théoremè de compacité. Compt. Rend. Sci., Paris, 256 (1963), 5042 - 5044.

BALASHOVA, G. S.

[1] Some extension theorems in Sobolev spaces of infinite order and nonhomogeneous boundary problems. (Russian) Dokl. Akad. Nauk USSR, 244, (1979) 6, 1294 - 1297.

[2] Some imbedding theorems of the spaces of infinitely differentiable functions. (Russian) Dokl. Akad. Nauk USSR, 247, (1979) 6, 1301 - 1304.

[3] Behaviour of the solutions of some boundary value problems with unbounded increase of the order of the equation. (Russian) Diff. Uravnenija, v. XVII, (1981) 2, 256 - 269.

[4] Imbedding theorems of some spaces of infinitely differentiable functions and nonlinear equations. (Russian) Dokl. Akad. Nauk USSR, 263, (1982) 5, 1037 - 1039.

[5] On some extension theorems in the spaces of infinitely differentiable functions. (Russian) Matemat. Sbornik, 118, (1982) 3, 371 - 385.

[6] Imbedding theorem of some Banach spaces of infinitely differentiable functions. (Russian) Matemat. Zametki, 35, (1984) 4, 505 - 516.

BESOV, O. V.; IL'IN, V. P.; NIKOL'SKIJ, S. M.

[1] Integral representations of functions and imbedding theorems. (Russian) Moscow; Nauka 1975.
(English translation: Scripta Series in Math., Washington, Halsted Press, New York, Toronto, London: V. H. Winston & Sons 1978/79).

BJORKEN, J.; DRELL, S.

[1] Relativistic quantum mechanics. McGraw-Hill Book Company 1964.

BOURBAKI, N.

[1] Espaces vectoriels topologiques. Ch. III - V, ASI 1129, 1955.

BROWDER, F. E.

[1] Nonlinear elliptic boundary value problems. Bull. Amer. Math. Soc., 69, 6, 862 - 874.

[2] Strongly nonlinear parabolic boundary value problems. Amer. Journ. Math., 86 (1964) 2, 339 - 357.

[3] Nonlinear monotone operators and convex sets in Banach spaces. Bull. Amer. Math. Soc., 71, (1965) 5, 780 - 785.

CHAN DYK VAN

[1] One boundary value problem for some degenerate nonlinear equations of infinite order. (Russian) Diff. Uravn., 14, (1978) 11, 2002 - 2011.

[2] Sobolev spaces of infinite order with weight in the strip and the solvability of some boundary value problems for some degenerate nonlinear equations. (Russian) Dokl. Akad. Nauk USSR, 240 (1978) 4, 794 - 797.

[3] Nontriviality of Sobolev spaces of infinite order with weight in the full Euclidean space and the solvability of some degenerate nonlinear equations. (Russian) Diff. Uravn., 15 (1979) 3, 507 - 513.

[4] Nontriviality of Sobolev-Orlicz spaces of infinite order in bounded domain of Euclidean space. (Russian) Dokl. Akad. Nauk USSR, 250 (1980) 6, 1331 - 1334.

[5] Résolubilité des problèmes aux limites pour équations non linéaires elliptiques d'ordre infini en classes de Sobolev-Orlicz. Compt. Rend. Ac. Sc., Paris, 290, ser. A (1980), 501 - 504.

[6] Nonlinear differential equations and functional spaces of infinite order. (Russian) Minsk: Isdatelstvo of Byelorussian State University, 1983, p. 118.

CLARKSON, J. A.

[1] Uniformly convex spaces. Trans. Amer. Math. Soc., 40 (1936), 396 - 414.

DAY, M. M.

[1] Normed linear spaces. Berlin, Göttingen, Heidelberg: Springer-Verlag 1958.

DUBINSKIJ, Ju. A.

[1] Weak convergence in nonlinear elliptic and parabolic equations. (Russian) Matem. Sborn., 67, (1965) 4, 609 - 642.

[2] Nonlinear elliptic and parabolic equations. (Russian) Itogi Nauki: Sovremennye Problemy Matem., 9, VINITI, Moscow (1976) 5 - 125.
(English translation: J. of Soviet Math. 12 (1979) 5, 475 - 554).

[3] Sobolev spaces of infinite order and the behaviour of solutions of some boundary-value problems with unbounded increase of the order of the equation. (Russian) Matem. Sborn., 98 (140), (1975) 2, 163 - 184.

[4] Nontriviality of Sobolev spaces of infinite order for the full Euclidean space and the torus. (Russian) Matem. Sborn., 100 (142), (1976) 3, 436 - 446.

[5] Traces of functions from Sobolev spaces of infinite order and inhomogeneous problems for nonlinear equations. (Russian) Matem. Sborn., 106 (148), (1978) 1, 66 - 84.

[6] Some problems of the theory of Sobolev spaces of infinite order and of nonlinear equations. Nonlinear Analysis, Functional Spaces and Applications (Proceedings of a Spring Scool held in Horni-Bradlo), Teubner-Texte zur Mathematik 19, 1978, 23 - 37.

[7] Some imbedding theorems of Sobolev spaces of infinite order. (Russian) Dokl. Akad. Nauk USSR, 242, (1978) 6, 1241 - 1244.

[8] Limits of Banach spaces. Imbedding theorems. Applications to Sobolev spaces of infinite order. Matem. Sborn., 110, (1979) 3, 428 - 439.

[9] On a method of solving partial differential equations. (Russian) Dokl. Akad. Nauk USSR, 258, (1981) 4, 780 - 784.

[10] On the theory of Cauchy problem for partial differential equations. (Russian) Dokl. Acad. Nauk USSR, 259, (1981) 4, 781 - 785.

[11] Algebra of differential operators of infinite order and pseudodifferential operators with analytic symbols. (Russian) Dokl. Akad. Nauk USSR, 264, (1982) 4, 807 - 812.

[12] Algebra of pseudodifferential operators with analytic symbols and applications to mathematical physics. (Russian) Uspehi Matemat. Nauk, 37, (1982) 5, 97 - 137.

[13] Pseudodifferential operators with complex arguments and applications. (Russian) Dokl. Akad. Nauk USSR, 268, (1983) 5, 1046 - 1050.

[14] Fourier transform of analytic functions. Complex method of Fourier. (Russian) Dokl. Akad. Nauk USSR, 275, (1984) 3, 533 - 536.

DUBINSKIJ, JU. A.; POHOŠAEV, S. I.

[1] A class of operators and the solvability of some quasilinear elliptic equations. (Russian) Matem. Sborn., 72 (114), (1967) 2, 226 - 236.

GELFAND, I. M.; ŠILOV, G. E.

[1] Generalized functions, v. I: Properties and operations. Fizmatgiz, Moscow 1958; (Russian) (English translation: New-York: Acad. Press 1964.

HÖRMANDER, L.

[1] Estimates for translation invariant operators in $L^p$ spaces. Acta Math., 104, (1960), 93 - 140.

KLENINA, L. I.

[1] Solvability of Cauchy-Dirichlet problem for some nonlinear elliptic equations of infinite order. (Russian) Dokl. Akad. Nauk USSR, 223, (1976) 1, 27 - 29.

KOBILOV, A. JA.

[1] Criterion of the nontriviality of Sobolev spaces of infinite order for an angular domain . (Russian) Doklady Akad. Uzbekskoi SSR, 5 (1980), 13 - 15.

[2] Nontriviality of some spaces of infinitely differentiable functions in angular domains and solvability of nonlinear elliptic equations. (Russian) Dokl. Akad. Nauk USSR, 266, (1982) 5, 1041 - 1044.

KONJAEV, JU. A.

[1] Asymptotic representation of the periodic solutions of some elliptic equations of order 2m in the process $m \to \infty$ . (Russian) Diff. Uravn., 14, (1978) 10, 1900 - 1902.

KUFNER, A.; JOHN, O.; FUČIK, S.

[1] Function spaces. Prague, 1977.

LATTES, R.; LIONS, J. L.

[1] Méthode de quasi-réversibilité et applications. Paris: Dunod 1967.

LAVRENTJEV, M. M.

[1] Some noncorrect problems in the mathematical physics. (Russian) Novosibirsk: Izdat. Sibirskogo Otdelenija Akad. Nauk USSR 1962.

LELONG, P.

[1] Sur une proprieté de quasianalyticité des fonctions de plusieurs variables. Compt. Rend. Acad. Sc., Paris, 232 (1951), 1178 - 1180.

LERAY, J.; LIONS, J. L.

[1] Quelques résultates de Višik sur les problèmes elliptiques non linéaires par les méthodes de Minty-Browder. Bull. Soc. Math. France, 93, (1965) 1, 97 - 107.

LEVIN, V. M.

[1] A connection between the mathematical expectations of intensity and of deformation in elastic micrononhomogeneous bodies. (Russian) Prikl. Mat. Mech., 35, (1971) 4, 744 - 750.

LIONS, J. L.

[1] Sur certaines équations paraboliques non lineaires. Bull. Soc. Math. France, 93, (1965) 1, 155 - 175.

[2] Equations differentielles opérationnelles et problemes aux limites. Grundl. d. math. Wiss. v. 111, Berlin: Springer 1961.

[3] Quelques méthodes de résolution des problêmes aux limites non linéaires. Paris: Dunod, Gauthier-Villars 1969.

MANDELBROJT, S.

[1] Séries adhérentes, régularisation des suites, applications. Paris: Gauthier-Villars 1952.

MARKUŠEVICH, A. I.

[1] Theory of analytic functions, v. 2. (Russian) Moscow: Nauka 1968.

MINTY, G.

[1] Monotone (nonlinear) operators in Hilbert spaces. Duke Math. Journ., 29, (1962) 3, 341 - 346.

[2] On a monotonicity method for the solution of nonlinear equation in Banach spaces. Proc. Nat. Acad. Sci. USA, 50, (1963) 6, 1051 - 1053.

NIKOL'SKIJ, S. M.

[1] Approximation of functions of several variables and imbedding theorems. (Russian) Moscow, Nauka, 1969. (English translation: Berlin, Heidelberg, New York: Springer-Verlag 1975).

NOVOŠILOV, V. V.

[1] A connection between the mathematical expectations of density tensors and of deformation tensors in some static isotropic elastic bodies. (Russian) Prikl. Math. Mech., 34, (1970) 1, 67 - 74.

PAUKŠTO, M. V.

[1] Mathematical questions of the theory of elasticity with a special reology. (Russian) Outline of dissertation, Leningrad, LGU, 1978.

ŠABAT, B. V.

[1] Introduction to complex analysis, v. I. (Russian) Moscow: Nauka 1975.

SAMAROV, K. L.

[1] On the fundamental solution for Schrödinger equation of relativistic free particle. (Russian) Dokl. Akad. Nauk USSR, 271, (1983) 2, 334 - 337.

SCHWARTZ, L.

[1] Théorie des distributions, t. I, II. Paris, 1951.

SOBOLEV, S. L.

[1] Some applications of functional analysis in mathematical physics. (Russian) Izdat. Sibirskogo Otdelenija Akad. Nauk USSR, Novosibirsk, 1962.

STEIN, E. M.; WEISS, G.

[1] Introduction to Fourier analysis on Euclidean spaces. Princeton: Princeton Univ. Press 1971.

TICHONOV, A. N.

[1] Solvability of some noncorrect problems and a method of regularization. (Russian) Dokl. Akad. Nauk USSR, 151, (1963) 3, 501 - 504.

UMAROV, S. R.

[1] Some spaces of infinite order and applications to operator equations. (Russian) Dokl. Akad. Nauk USSR, 275, (1984) 2, 313 - 317.

VIŠIK, M. I.

[1] Quasilinear strongly elliptic systems having the divergence form. (Russian) Trudy Moscovskogo Math. Obšestva, 12, (1963), 125 - 184.

[2] Solvability of boundary value problems for some parabolic quasilinear equations. (Russian) Matem. Sborn. 59 (dopolnielny) (1962), 289 - 325.

VLADIMIROV, V. S.

[1] Equations of mathematical physics. (Russian) Moscow, Nauka 1967.

VOLEVICH, L. R.; PANEJAH, B. P.

[1] Some spaces of the generalized functions and imbedding theorems. (Russian) Uspehi Matem. Nauk, XX, (121) (1965) 1, 3 - 74.

ZYGMUND, A.

[1] Trigonometric series, v. I. II. Cambridge: Cambridge Univ. Press 1959.